Kotlin Programming Cookbook

Explore more than 100 recipes that show how to build robust mobile and web applications with Kotlin, Spring Boot, and Android

Aanand Shekhar Roy
Rashi Karanpuria

BIRMINGHAM - MUMBAI

Kotlin Programming Cookbook

Commissioning Editor: Merint Mathew
Acquisition Editor: Isha Raval
Content Development Editor: Onkar Wani
Technical Editor: Sachin Sunilkumar
Copy Editor: Shaila Kusanale
Project Coordinator: Devanshi Doshi
Proofreader: Safis Editing
Indexer: Rekha Nair
Graphics: Jason Monteiro
Production Coordinator: Nilesh Mohite

First published: January 2018

Production reference: 1230118

Published by Packt Publishing Ltd.
Livery Place
35 Livery Street
Birmingham
B3 2PB, UK.

ISBN 978-1-78847-214-2

www.packtpub.com

Contributors

About the authors

Aanand Shekhar Roy is a freelance Android developer. His mobile engineering career includes working for various startups and companies, such as Netherlands-based ConceptOffice and U.S.-based startups Consciously.Life and NextFan-FantasyIndyCar. He is also a mentor for the Associate Android Developer Fast Track program by Google on Udacity and mentors a team of 20 software developers. He is also a writer at AndroidPub.

Acknowledgement is probably that part of the book you least have interest in. Simply because every time you open it, it's either thanking someone else who you don't know. Not this time, this book is dedicated to only learners like you, simply because it wouldn't have come to life without you wanting to read it.

Rashi Karanpuria is a professional Android developer. She started her career by working on a wallet app for a New Zealand based Fintech startup. She has developed projects in various domains such as IOT, AdTech, Business, Social, and live streaming apps. Over the years, she has acquired a deep understanding of the user experience needed to develop a successful mobile application. She combines her design skills with scalable and maintainable code using best practices to create quality products.

I would like to thank each & every member of the developer community all across the world. This is the only community that I know of to give ideas and resources so freely without expecting anything in return. Did you know Kotlin is open source? This giving nature of our developer community is why I am who I am today. Thanks.

`mapt.io`

Mapt is an online digital library that gives you full access to over 5,000 books and videos, as well as industry leading tools to help you plan your personal development and advance your career. For more information, please visit our website.

Why subscribe?

- Spend less time learning and more time coding with practical eBooks and Videos from over 4,000 industry professionals

- Improve your learning with Skill Plans built especially for you

- Get a free eBook or video every month

- Mapt is fully searchable

- Copy and paste, print, and bookmark content

PacktPub.com

Did you know that Packt offers eBook versions of every book published, with PDF and ePub files available? You can upgrade to the eBook version at `www.PacktPub.com` and as a print book customer, you are entitled to a discount on the eBook copy. Get in touch with us at `service@packtpub.com` for more details.

At `www.PacktPub.com`, you can also read a collection of free technical articles, sign up for a range of free newsletters, and receive exclusive discounts and offers on Packt books and eBooks.

About the reviewer

Mitchell Wong Ho was born in Johannesburg, South Africa, where he completed his national diploma in electrical engineering. Mitchell's software development career started on embedded systems and then moved to Microsoft desktop/server applications. Mitchell has been programming in Java since 2000 on J2ME, JEE, desktop, and Android applications, and has more recently been advocating Kotlin for Android.

Packt is searching for authors like you

If you're interested in becoming an author for Packt, please visit `authors.packtpub.com` and apply today. We have worked with thousands of developers and tech professionals, just like you, to help them share their insight with the global tech community. You can make a general application, apply for a specific hot topic that we are recruiting an author for, or submit your own idea.

Table of Contents

Preface

Kotlin Cookbook will be a go-to guide to problems new Kotlin developers get stuck with. Along with that, Kotlin Cookbook will also help developers learn handy tricks and concepts that they may need while coding. This book will also help developers uncover the full potential of an amazing programming language, that is, Kotlin.

The book starts with an overview of Kotlin and moves on to some great simple concepts and features that Kotlin offers. From there, it will move on to OOP fundamentals and creating simple Android applications. Next will be the recipes for more complicated concepts such as networking, database, architectures, file io, and testing. It will also cover some great features of Anko that really eases out some complicated concepts in Android development, making it faster and more fun. Last will be some miscellaneous but extremely useful recipes that developers might need from time to time.

Who this book is for

This book is targeted at Kotlin beginners who know Android and Java Development and who have a good knowledge level and understanding of the Android development cycle. The readers are familiar with the concepts of Android development and understand the needs of testing their code. They want to learn efficient Kotlin techniques in order to make the existing Android development process more efficient and fun. This *is not* an introductory book for Kotlin, and it assumes basic familiarity with Kotlin. This book aims at helping developers solve issues they are stuck with while working with Kotlin.

What this book covers

Chapter 1, *Installation and Working with Environment*, walks you through starting with a Kotlin project. We will also introduce you to the Gradle build system and help you in setting up your development environment.

Chapter 2, *Control Flow*, includes recipes for control flow in Kotlin. Kotlin has brought a lot of power to old control flows, as you can now use them as an expression. Kotlin has also introduced a powerful "when", which is basically Java's "switch" improvements.

Chapter 3, *Classes and Objects*, says that classes and objects are inevitable parts of object-oriented programming. This chapter will include the solutions and examples of real-world problems faced by developers and how Kotlin solves it. This chapter will also lay the foundation to the upcoming chapter OOPS programming with Kotlin.

Chapter 4, *Functions*, informs that functions are inevitable parts of object-oriented programming. This chapter will include the solutions and examples of real-world problems faced by developers and how Kotlin solves it.

Chapter 5, *Object-Oriented Programming*, builds upon the learning of Chapter 3, *Classes and Objects*, and will include recipes that help in OOP.

Chapter 6, *Collections Framework*, presents the recipes that will explore the full potential of Collection framework in Kotlin.

Chapter 7, *Handling File Operations in Kotlin*, covers recipes about basic I/O and File I/O.

Chapter 8, *Anko Commons and Extension Function*, contains recipes on how to use the Anko library of Kotlin for efficient and quick Android development.

Chapter 9, *Anko Layouts*, has recipes on how to use the Anko library of Kotlin for efficient and quick Android development.

Chapter 10, *Databases and Dependency Injection*, dives into recipes to work with SQLite databases in Android.

Chapter 11, *Networking and Concurrency*, discusses recipes that will help developers make network calls and fetch data over a network.

Chapter 12, *Lambdas and Delegates*, uncovers some of the best (and difficult) features of Kotlin, that is, Lambdas and Delegates. This contains recipes to help the developers get started with them.

Chapter 13, *Testing*, outlines concepts on writing tests in Kotlin while touching Unit tests, integration tests, instrumentation, and acceptance tests.

Chapter 14, *Web Services with Kotlin*, helps developers write web services using Kotlin language.

To get the most out of this book

This book assumes familiarity with Java and Android development. This is not an introductory book for learning Kotlin. Readers must have used Android studio because many of the recipes will be focused toward Android development.

Download the example code files

You can download the example code files for this book from your account at `www.packtpub.com`. If you purchased this book elsewhere, you can visit `www.packtpub.com/support` and register to have the files emailed directly to you.

You can download the code files by following these steps:

1. Log in or register at `www.packtpub.com`.
2. Select the **SUPPORT** tab.
3. Click on **Code Downloads & Errata**.
4. Enter the name of the book in the **Search** box and follow the onscreen instructions.

Once the file is downloaded, please make sure that you unzip or extract the folder using the latest version of:

- WinRAR/7-Zip for Windows
- Zipeg/iZip/UnRarX for Mac
- 7-Zip/PeaZip for Linux

The code bundle for the book is also hosted on GitLab at `https://gitlab.com/users/aanandshekharroy/projects`. We also have other code bundles from our rich catalog of books and videos available at `https://github.com/PacktPublishing/`. Check them out!

Conventions used

There are a number of text conventions used throughout this book.

`CodeInText`: Indicates code words in text, database table names, folder names, filenames, file extensions, pathnames, dummy URLs, user input, and Twitter handles. Here is an example: "The corresponding `sourceSets` property should be updated if not using the default convention."

A block of code is set as follows:

```
sourceSets {
    main.kotlin.srcDirs += 'src/main/myKotlin'
    main.java.srcDirs += 'src/main/myJava'
}
```

When we wish to draw your attention to a particular part of a code block, the relevant lines or items are set in bold:

```
sourceSets {
    main.java.srcDirs += 'src/main/kotlin/'
}
```

Any command-line input or output is written as follows:

```
$ kotlinc hello.kt -include-runtime -d hello.jar.
$ java -jar hello.jar
```

Bold: Indicates a new term, an important word, or words that you see onscreen. For example, words in menus or dialog boxes appear in the text like this. Here is an example: "In the **Select Deployment Target** window, select your device, and click on **OK**."

Warnings or important notes appear like this.

Tips and tricks appear like this.

Sections

In this book, you will find several headings that appear frequently (*Getting ready, How to do it..., How it works..., There's more...,* and *See also*).

To give clear instructions on how to complete a recipe, use these sections as follows:

Getting ready

This section tells you what to expect in the recipe and describes how to set up any software or any preliminary settings required for the recipe.

How to do it...

This section contains the steps required to follow the recipe.

How it works...

This section usually consists of a detailed explanation of what happened in the previous section.

There's more...

This section consists of additional information about the recipe in order to make you more knowledgeable about the recipe.

See also

This section provides helpful links to other useful information for the recipe.

Get in touch

Feedback from our readers is always welcome.

General feedback: Email `feedback@packtpub.com` and mention the book title in the subject of your message. If you have questions about any aspect of this book, please email us at `questions@packtpub.com`.

Errata: Although we have taken every care to ensure the accuracy of our content, mistakes do happen. If you have found a mistake in this book, we would be grateful if you would report this to us. Please visit www.packtpub.com/submit-errata, selecting your book, clicking on the Errata Submission Form link, and entering the details.

Piracy: If you come across any illegal copies of our works in any form on the internet, we would be grateful if you would provide us with the location address or website name. Please contact us at copyright@packtpub.com with a link to the material.

If you are interested in becoming an author: If there is a topic that you have expertise in and you are interested in either writing or contributing to a book, please visit authors.packtpub.com.

Reviews

Please leave a review. Once you have read and used this book, why not leave a review on the site that you purchased it from? Potential readers can then see and use your unbiased opinion to make purchase decisions, we at Packt can understand what you think about our products, and our authors can see your feedback on their book. Thank you!

For more information about Packt, please visit packtpub.com.

1
Installation and Working with Environment

The following recipes will be covered in this chapter:

- Creating Kotlin Android project
- How to use Gradle to run Kotlin code
- How to run a Kotlin compiled class
- How to build a self-executable jar with Gradle and Kotlin
- Reading console input in Kotlin
- Converting Java code to Kotlin and vice versa
- How to write an idiomatic logger with Kotlin
- Escaping for Java identifiers that are keywords in Kotlin
- Disambiguating using the "as" keyword to locally rename the clashing entity
- Doing bit manipulations in Kotlin
- Parsing String to Long, Double, or Int
- Using String templates in Kotlin

Introduction

Android apps are a fascinating piece of technology. The apps developed on Android have worldwide appeal and audience. However, that has posed serious challenges for developers. The challenge is with updating APIs, platforms, and varied device capabilities. For example, if you are an Android developer, you have to rely on Java 6 if you want to support all API levels in Android. Java 6 is obsolete now, so much so that even its successor, Java 7, is kind of obsolete today. There was a great need for modern language for Android, which has built a trillion dollar industry around it and has influenced billions of lives. True, we have Java 8 now, but we can only use it if we are developing Android apps for API level 24 and above. However, that's equivalent to targeting only 9% of Android devices as of 2017; clearly, this isn't the way to go.

All is not lost though, and thanks to the JVM, we can write Android apps using any language that produces JVM compatible bytecode on compilation. So theoretically, we can use Clojure, Groovy, Scala, and Kotlin, but Kotlin is the best alternative among all, why? It's because in April 2017, Google announced Kotlin as an official language for Android development.

Some of the biggest tech companies such as Pinterest, Uber, Atlassian, Coursera, and Evernote are now using Kotlin in their Android apps. This wide adoption by them already speaks a huge volume for Kotlin. The 100% interoperability with Android and Java has helped Kotlin in its adoption. Kotlin is much easier to work with than Java and, apart from Android apps, you can also build web-apps with it. So, this chapter will introduce you to Kotlin and help you get started with this awesome piece of technology.

In this chapter, we will first see how to set up the environment to begin working with Kotlin.

Creating Kotlin Android project

Getting started with Kotlin is really easy, especially after Google has added official support for the language. You can use Kotlin directly with Android Studio 3. Android Studio 3 is still in Beta version at the time of writing this book. The best thing about using Kotlin for Android is that it is interoperable with your existing code, be it Java or C++. While working with Kotlin, you will realize that code in Kotlin is concise, extensible, and powerful. It really makes Android development more fun. Let's see how we can start working in Kotlin by first creating a Kotlin project in Android Studio 3.

Getting ready

To get started with this recipe, you will need Android Studio installed on your computer. Android Studio has both Android SDK and Android Virtual device in it. Ensure that you have Java Development Kit installed on your system. You will need an android phone or Emulator for debugging your project. You will also need at least one Android Virtual Device installed, of your desired specifications if you are not using an Android phone.

So basically, here's the checklist of the things that need to be installed before you move on to the next section:

- Java Development Kit (use the latest)
- Android Studio 3+
- Android phone or emulator

How to do it...

Creating a project in Android Studio is very simple and to create it in Kotlin just requires one extra click. Here's a step-by-step process of doing it:

1. In Android Studio, in the menu, click on **File** | **New** | **New Project**. Alternatively, if you've just opened Android Studio and see the Welcome to Android Studio window, click on **Start a new Android Studio project**.

2. In the wizard, add your Application name and Company domain, and simply check the box that says **Include Kotlin support**. Click on **Next**:

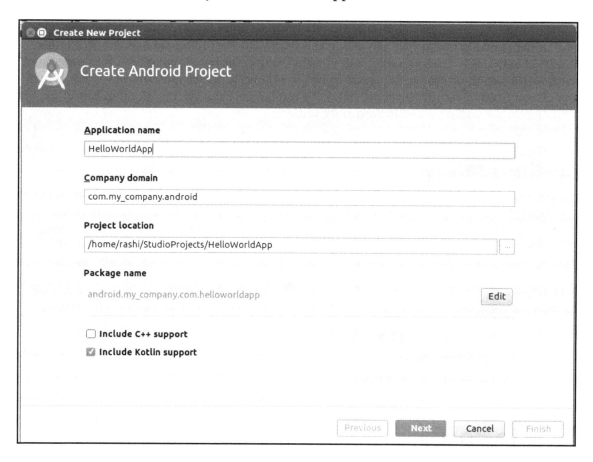

3. On the next screen, you will be asked to choose your target devices and the minimum SDK support. So basically, it asks things like, "Do you want the application to run on both phone and android wear?" and "Do you want to support from Jelly Bean up or KitKat and up?":

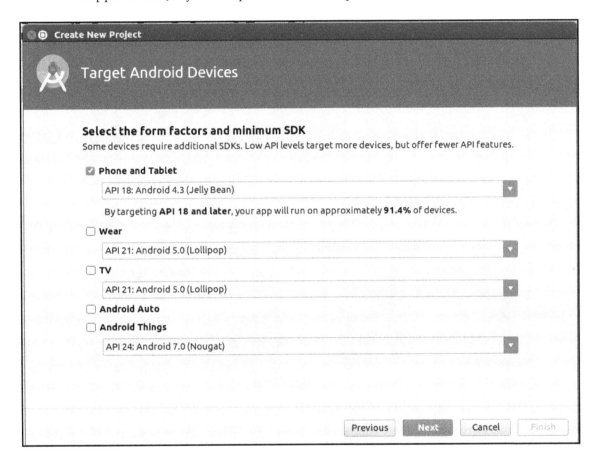

4. On the next screen, you will be prompted to **Add an Activity** to the project. You can also skip this step and add an activity later, but for now, just click on a **Basic Activity** and click on **Next**. If you have also chosen **Wear** or any other option, you will be prompted to add activity for those components as well:

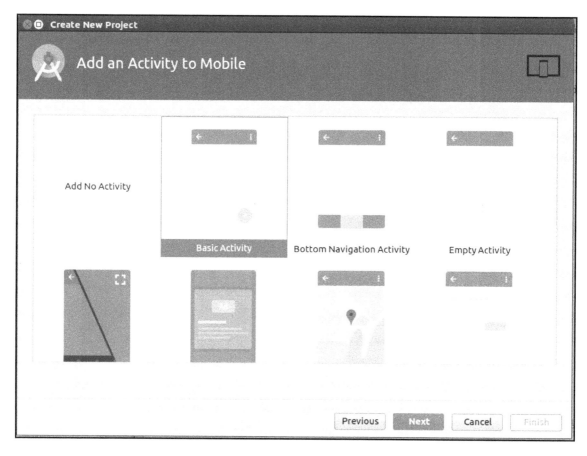

5. Next, you will be prompted to **Configure the Activity** you added. Basically, what you have to do is to provide **Activity Name, Layout Name,** and **Title**. After this, click on **Finish**, because you are done with creating your first project in Kotlin.

6. Run project on your device: You need to follow these steps:
    1. Connect your device to your development machine with a USB cable.
    2. Enable USB debugging on your device by going to **Settings | Developer options**.

On Android 4.2 and newer, Developer options are hidden by default. To make it available, go to **Settings** | **About phone** and tap on **Build number** seven times. Return to the previous screen to find **Developer options**.

Now in your Android Studio, click on the **app** module in the **Project** window and then select **Run** (or click on **Run** in the toolbar).

In the **Select Deployment Target** window, select your device, and click on **OK**. After a while, you will see the application running on your mobile or an emulator.

There's more...

After clicking on the **Finish** button in the **Create New Project** window, Android Studio will configure things and create your project. If you added an activity as mentioned in Step 4, you will be greeted with the boilerplate code of the activity. It looks something like this:

```
File Edit View Navigate Code Analyze Refactor Build Run Tools VCS Window Help

orldApp ⟩ app ⟩ src ⟩ main ⟩ java ⟩ android ⟩ my_company ⟩ com ⟩ helloworldapp ⟩ HelloWorldActivity.kt

  Android                        HelloWorldActivity.kt   content_hello_world.xml

    app
    Gradle Scripts
                        package android.my_company.com.helloworldapp

                        import ...

                        class HelloWorldActivity : AppCompatActivity() {

                            override fun onCreate(savedInstanceState: Bundle?) {
                                super.onCreate(savedInstanceState)
                                setContentView(R.layout.activity_hello_world)
                                setSupportActionBar(toolbar)

                                fab.setOnClickListener { view ->
                                    Snackbar.make(view, "Replace with your own action", Snackbar.LENGTH_LONG)
                                        .setAction("Action", null).show()
                                }
                            }

                            override fun onCreateOptionsMenu(menu: Menu): Boolean {
                                // Inflate the menu; this adds items to the action bar if it is present.
                                menuInflater.inflate(R.menu.menu_hello_world, menu)
                                return true
                            }

                            override fun onOptionsItemSelected(item: MenuItem): Boolean {
                                // Handle action bar item clicks here. The action bar will
                                // automatically handle clicks on the Home/Up button, so long
                                // as you specify a parent activity in AndroidManifest.xml
                                return when (item.itemId) {
                                    R.id.action_settings -> true
                                    else -> super.onOptionsItemSelected(item)
                                }
                            }
                        }

  Terminal   6: Logcat   Android Profiler   0: Messages   TODO                          Event Log   Gradle Console
  Gradle build finished in 6s 296ms (57 minutes ago)                                     1:1   LF:  UTF-8
```

How to use Gradle to run Kotlin code

Gradle has now become the de facto build tool for Android, and it is very powerful. It's great for automating tasks without compromising on maintainability, usability, flexibility, extensibility, or performance. In this recipe, we will see how to use Gradle to run Kotlin code.

Getting ready

We will be using IntelliJ IDEA because it provides great integration of Gradle with Kotlin, and it is a really great IDE to work on. You can also use Android Studio for it.

How to do it...

In the following steps, we will be creating a Kotlin project with the Gradle build system. First, we will select the **Create New Project** option from the menu. Then, follow these steps:

1. Create the project with the **Gradle** build system:

2. After you have created the project, you will have the `build.gradle` file, which will look something like the following:

```
version '1.0-SNAPSHOT'

buildscript {
  ext.kotlin_version = '1.1.4-3'

  repositories {
      mavenCentral()
  }
  dependencies {
      classpath "org.jetbrains.kotlin:kotlin-gradle-
plugin:$kotlin_version"
  }
}

apply plugin: 'java'
apply plugin: 'kotlin'

sourceCompatibility = 1.8

repositories {
  mavenCentral()
}

dependencies {
  compile "org.jetbrains.kotlin:kotlin-stdlib-jre8:$kotlin_version"
  testCompile group: 'junit', name: 'junit', version: '4.12'
}

compileKotlin {
  kotlinOptions.jvmTarget = "1.8"
}
compileTestKotlin {
  kotlinOptions.jvmTarget = "1.8"
}
```

3. Now we will create a `HelloWorld` class, which will have a simple main function:

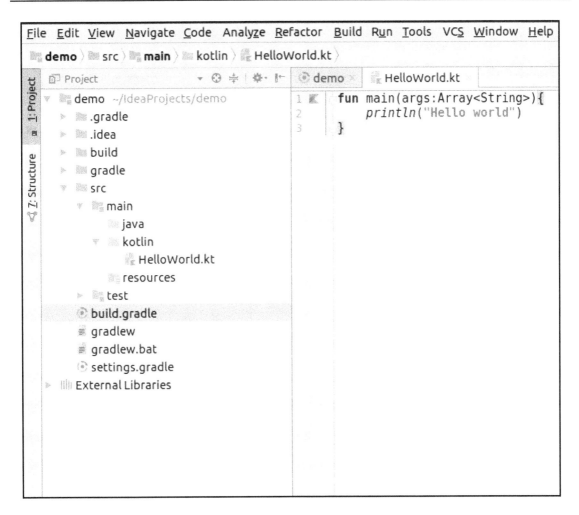

4. Now, it would be really cool to run this code directly. To do so, we will use the `gradle run` command. However, before that, we need to enable the application plugin, which will allow us to directly run this code. We need to add two lines in the `build.gradle` file to set it up:

```
apply plugin: 'application'
mainClassName = "HelloWorldKt"
```

5. After this, you can type `gradle run` in the terminal to execute this file, and you will see the output of the method, as shown:

There's more...

The default structure of the project, when you create a new project in IntelliJ, is as illustrated:

```
project
    - src
        - main (root)
            - kotlin
            - java
```

If you want to have a different structure of the project, you should declare it in `build.gradle`. You can do it by adding the following lines in `build.gradle`.

The corresponding `sourceSets` property should be updated if not using the default convention:

```
sourceSets {
    main.kotlin.srcDirs += 'src/main/myKotlin'
    main.java.srcDirs += 'src/main/myJava'
}
```

Though you can keep Kotlin and Java files under the same package, it's a good practice to keep them separated.

See also

Check out the *How to build a self-executable jar with Gradle and Kotlin* recipe in this chapter.

How to run a Kotlin compiled class

Working with the command-line compiler for any language is one of the first steps to get a better understanding of the language, and this knowledge comes handy at a lot of times. In this recipe, we will run a Kotlin program using the command line, and we will also play a bit with the interactive shell of Kotlin.

Getting ready

To be able to perform this recipe, you need a Kotlin compiler installed on your development machine. Every Kotlin release ships with a standalone compiler. You can find the latest release at `https://github.com/JetBrains/kotlin/releases`.

To manually install the compiler, unzip the standalone compiler into a directory and optionally, add the bin directory to the system path. The bin directory contains the scripts needed to compile and run Kotlin on Windows, OS X, and Linux.

How to do it...

Now we are ready to run our first program using the command line. First, we will create a simple application that displays Hello World! and then compile it:

1. Create a file with the name `hello.kt` and add the following lines of code in that file:

```
fun main(args: Array<String>) {
    println("Hello, World!")
}
```

2. Now we compile the file using the following command:

```
$ kotlinc hello.kt -include-runtime -d hello.jar
```

3. Now we run the application using the following command:

```
$ java -jar hello.jar
```

4. Suppose you want to create a library that can be used with other Kotlin applications; we can simply compile the Kotlin application in question into `.jar` executable without the `-include-runtime` option, that is, the new command will be as follows:

```
$ kotlinc hello.kt -d hello.jar
```

5. Now, let's check out the Kotlin interactive shell. Just run the Kotlin compiler without any parameters to have an interactive shell. Here's how it looks:

```
rashi ~ $ kotlinc
Welcome to Kotlin version 1.1.50 (JRE 1.8.0_131-8u131-b11-2ubuntu1.16.04.3-b11)
Type :help for help, :quit for quit
>>>
```

Hopefully, you must have noticed the information I am always guilty of ignoring, that is, the command to quit interactive shell is :quit and for help, it is :help.

You can run any valid Kotlin code in the interactive shell. For example, try some of the following commands:

- `3*2+(55/5)`
- `println("yo")`
- `println("check this out ${3+4}")`

Here's a screenshot of running the preceding code:

```
  ⊗ ⊝ ⊡   /bin/bash
                        /bin/bash 79x28



rashi ~ $ kotlinc
Welcome to Kotlin version 1.1.50 (JRE 1.8.0_131-8u131-b11-2ubuntu1.16.04.3-b11)
Type :help for help, :quit for quit
>>> 3*2+(55/5)
17
>>> println("yo")
yo
>>> println("check this out ${3+4}")
check this out 7
>>>
```

How it works...

The -include-runtime option makes the resulting .jar file self-contained and runnable by including the Kotlin runtime library in it. Then, we use Java to run the .jar file generated.

The -d option in the command indicates what we want the output of the compiler to be called and maybe either a directory name for class files or a .jar filename.

There's more...

Kotlin can also be used for writing shell scripts. A shell script has top-level executable code.

Kotlin script files have the `.kts` extension as opposed to the usual `.kt` for Kotlin applications.

To run a script file, just pass the `-script` option to the compiler:

```
$ kotlinc -script kotlin_script_file_example.kts
```

How to build a self-executable JAR with Gradle and Kotlin

Kotlin is great for creating small command-line utilities, which can be packaged and distributed as normal JAR files. In this recipe, we will see how to do it using Gradle build system. Gradle build system is one of the most sophisticated build systems out there. It is the default build tool for Android and is designed to ease scripting of complex, multilanguage builds with a lot of dependencies (typical of big projects). It achieves the goal of automating your project without compromising on maintainability, usability, flexibility, extensibility, or performance. We will be using Gradle build system to create a self-extracting JAR file. This JAR file can be distributed to and run on any platform supporting Java.

Getting ready

You need an IDE (preferably IntelliJ or Android Studio), and you need to tell it where your Kotlin files are present. You can do so by specifying it in the `build.gradle` file by adding the following:

```
sourceSets {
    main.java.srcDirs += 'src/main/kotlin/'
}
```

The preceding lines are required if you have your Kotlin files separated from Java packages. This is optional, and you can continue working with Kotlin files under Java packages, but it's a good practice to keep them separated.

We'll be creating a very simple function that just prints `Hello World!` when executed. Since it'll be a simple function, I am just adding it as a top-level `main()` function.

How to do it...

Let's go through these steps, with which we can create a self-executable JAR:

1. We'll create a simple class `HelloWorld.kt` having the main function, which just prints out "Hello world!":

```
fun main(args:Array<String>){
    println("Hello world")
}
```

2. Now we need to configure a `jar` task, which Gradle build goes through to inform it of our entry to our project. In a Java project, this will be the path to the class where our `main()` function resides, so you will need to add this `jar` task in `build.gradle`:

```
jar {
    manifest {
        attributes 'Main-Class': 'HelloWorldKt'
    }
    from { configurations.compile.collect { it.isDirectory() ? it :
zipTree(it) } }
}
```

3. After adding the preceding snippet to `build.gradle`, you need to run the following gradle command to create the jar file:

./gradlew clean jar

4. The created jar file can be found in the `build/libs` folder. Now you can just run the `java -jar demo.jar` command to run the JAR file.

After you do that, you can see the output in the console:

```
theseus libs $ java -jar demo-1.0-SNAPSHOT.jar
Hello world
theseus libs $
```

How it works...

To make self-executable JARs, we need a manifest file called `MANIFEST.MF` in the `META-INF` directory. For our purposes here, we just need to specify the name of the Java class that contains the Java-based extractor program's `main()` method.

One might argue that even though we don't have top-level class declaration, we are specifying it as `HelloWorldKt` in the code for the jar task:

```
manifest {
    attributes 'Main-Class': 'HelloWorldKt'
}
```

The reason for putting the preceding code block in the jar task is that Kotlin compiler adds all top-level functions to respective classes for back-compatibility with JVM. So, the class generated by Kotlin compiler will have the filename, plus the `Kt` suffix, which makes it `HelloWorldKt`.

Also, the reason we added `from { configurations.compile.collect {` `it.isDirectory() ? it : zipTree(it) } }` in jar task is because we want Gradle to copy all of a JAR's dependencies. The reason for doing so is that, by default, when Gradle (as well as Maven) packs some Java class files into a JAR file, it assumes that this JAR file will be referenced by an application, where all of its dependencies are also accessible in the classpath of the loading application. So, by specifying the preceding lines in jar task, we are telling gradle to take all of this JAR's referenced dependencies and copy them as part of the JAR itself. In the Java community, this is known as a **fat JAR**. In a fat JAR, all the dependencies end up within the classpath of the loading application, so the code can be executed without problems. The only downside to creating fat JARs is their growing file size (which kind of explains the name), though it is not a big concern in most situations.

Reading console input in Kotlin

In many applications, user interaction is a very important part, and the most basic way of doing that is reading input entered by the user and giving output based on it. In this recipe, we will understand different ways of reading input and also provide output in the console.

Getting ready

You need to install a preferred development environment that compiles and runs Kotlin. You can also use the command line to compile and run your Kotlin code, for which you need Kotlin compiler installed along with JDK.

How to do it...

Let's go through the following steps by which we can read console input in Kotlin:

1. We will start simple and move to more advanced logic as we move forward. First, let's start with simply printing a line as output in the console:

```
println("Just a line")
```

2. Now we will try to take String input from the console and output it again:

```
println("Input your first name")
var first_name = readLine()
println("Your first name: $first_name")
```

3. Okay, how about we repeat the process with Int:

```
println("Hi $first_name, let us have a quick math test. Enter two
numbers separated by space.")
val (a, b) = readLine()!!.split(' ').map(String::toInt)
println("$a + $b = ${a+b}")
```

4. Now, let's try a complicated code and then start with the explanations:

```
fun main(args: Array<String>) {
   println("Input your first name")
   var first_name = readLine()
   println("Input your last name")
   var last_name = readLine()
   println("Hi $first_name $last_name, let us have a quick math
test. Enter two numbers separated by space.")
   val (a, b) = readLine()!!.split(' ').map(String::toInt)
  println("what is $a + $b ?")
  println("Your answer is ${if (readLine()!!.toInt() == (a+b))
"correct" else "incorrect"}")
   println("Correct answer = ${a+b}")
 println("what is $a * $b ?")
   println("Your answer is ${if (readLine()!!.toInt() == (a*b))
"correct" else "incorrect"}")
```

```
        println("Correct answer = ${a*b}")
        println("Thanks for participating :)")
    }
```

Here's a screenshot of compiling and running the preceding code:

How it works...

Let's try to understand the methods by which we were able to read input in Kotlin.

Behind the scenes, `Kotlin.io` uses `java.io` for the input-output. So `println` is basically `System.out.println`, but with additional power by Kotlin to use String templates and `inline` functions, which makes writing extremely simple and concise.

This is a part of the actual code from Kotlin `stdlib` used for Console IO:

```
/** Prints the given message and newline to the standard output
stream. */
@kotlin.internal.InlineOnly
public inline fun println(message: Any?) {
    System.out.println(message)
}
```

Converting Java code to Kotlin and vice versa

The best part about Kotlin is its interoperability with Java. Also, with IntelliJ-based IDE, we can directly convert our Java code to Kotlin. In this recipe, we will see how to do it.

Getting ready

This recipe needs IntelliJ-based IDE installed, which compiles and runs Kotlin and Java.

How to do it...

Let's see the steps to convert a Kotlin file to a Java file:

1. In your IntelliJ IDE, open the Java file that you want to convert to Kotlin.
2. Note that it has a `.java` extension. Now, in the main menu, click on **Code** menu and choose the **Convert Java File to Kotlin File** option. Your Java file will be converted into Kotlin, and the extension will now be `.kt`.

Shown here is an example of a Java file:

```
TestClass.java
    package android.my_company.com.helloworldapp;

    public class TestClass {
        String name;
        String year;
        int roll_number;

        public TestClass(String name, String year, int roll_number) {
            this.name = name;
            this.year = year;
            this.roll_number = roll_number;
        }

        public String getName() { return name; }

        public void setName(String name) { this.name = name; }

        public String getYear() { return year; }

        public void setYear(String year) { this.year = year; }

        public int getRoll_number() { return roll_number; }

        public void setRoll_number(int roll_number) { this.roll_number = roll_number; }
    }
```

After converting to Kotlin, this is what we have:

```
TestClass.kt
    package android.my_company.com.helloworldapp

    class TestClass(var name: String, var year: String, var roll_number: Int)
```

3. A Kotlin file can be converted into Java, but it's better if you can avoid it or find an alternative way to do it. If you have to absolutely convert your Kotlin code to Java, click on **Tools | Kotlin | Show Kotlin Bytecode** in the menu:

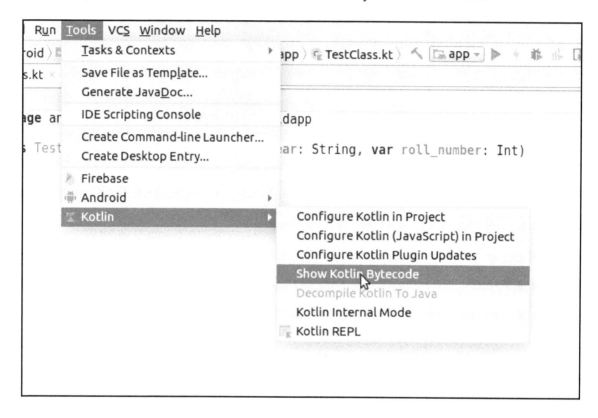

4. After clicking on **Show Kotlin Bytecode**, a window will open with the title **Kotlin Bytecode**:

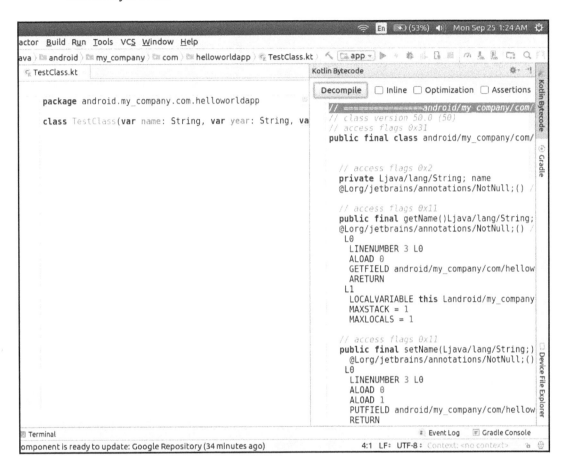

5. Click on **Decompile** and a `.java` file will be generated, containing a decompiled Java bytecode from Kotlin code:

```
TestClass.kt        TestClass.decompiled.java

    package android.my_company.com.helloworldapp;

    import ...

    @Metadata(
        mv = {1, 1, 7},
        bv = {1, 0, 2},
        k = 1,
        d1 = {"\u0000\u001a\n\u0002\u0018\u0002\n\u0002\u0010\u0000\n\u0000\n\u0002\u0010\u000e\n\u0002
        d2 = {"Landroid/my_company/com/helloworldapp/TestClass;", "", "name", "", "year", "roll_number"
    )
    public final class TestClass {
        @NotNull
        private String name;
        @NotNull
        private String year;
        private int roll_number;

        @NotNull
        public final String getName() { return this.name; }

        public final void setName(@NotNull String var1) {
            Intrinsics.checkParameterIsNotNull(var1,  paramName: "<set-?>");
            this.name = var1;
        }

        @NotNull
        public final String getYear() { return this.year; }

        public final void setYear(@NotNull String var1) {
            Intrinsics.checkParameterIsNotNull(var1,  paramName: "<set-?>");
            this.year = var1;
        }
```

Yes, it has a lot of unnecessary code that was not present in the original Java code, but that is the case with decompiled bytecode. At the moment, this is the only way to convert Kotlin code to Java. Copy the decompiled file into a `.java` file and remove the unnecessary code.

How it works...

Kotlin is a statically-typed programming language that works on Java Virtual Machine and compiles into JVM compatible bytecode. This is the reason we can convert Java code to Kotlin and mix Java and Kotlin code together. This is also the reason why you can, in a way, get Java code back from Kotlin (although the output is not completely desired).

How to write an idiomatic logger in Kotlin

Kotlin has some great powerful features packed in it that we should be making use of to improve our code. This involves rethinking on our old best practices of coding. Many of our old coding practices can be replaced by better alternatives from Kotlin. One of them is how we write our logger. Though there are a lot of libraries out there that provide logging functionality, we will try to create our own logger in this recipe, just by using idiomatic Kotlin.

Getting ready

We will be using IntelliJ IDE to write and execute our code.

How to do it...

Let's go through the given steps to create an idiomatic logger in Kotlin:

1. First, let's see how it was done in Java. In Java, SLF4J is used and considered de-facto, so much that logging seems like a solved problem in Java language. Here's what a Java implementation would look like:

```
private static final Logger logger =
LoggerFactory.getLogger(CurrentClass.class);
...
logger.info("Hi, {}", name);
```

2. It also works fine with Kotlin, obviously with minor modifications:

```
val logger = LoggerFactory.getLogger(CurrentClass::class)
...
logger.info("Hi, {}", name)
```

However, apart from this, we can utilize the power of Kotlin using **Delegates** for the logger. In this case, we will be creating the logger using the `lazy` keyword. This way, we will create the object only when we access it. Delegates are a great way to postpone object creation until we use it. This improves startup time (which is much needed and appreciated in Android). So let us explore a method using lazy delegates in Kotlin:

1. We'll use `java.util.Logging` internally, but this works for any Logging library of your choice. So let's use the Kotlin's lazy delegate to get our logger:

```
public fun <R : Any> R.logger(): Lazy<Logger> {
    return lazy { Logger.getLogger(this.javaClass.name) }
}
```

2. Now in our class, we can simply call the method to get our logger and use it:

```
class SomeClass {
  companion object { val log by logger() }

  fun do_something() {
      log.info("Did Something")
  }
}
```

When you run the code, you can see the following output:

```
Sep 25, 2017 10:49:00 PM packageA.SomeClass do_something
INFO: Did Something
```

So, as we can see in the output, we get the class name and method name too (if you are accessing logger inside a method).

How it works...

Here, one thing to note is that we have put our logger inside a companion object. The reason for this is quite straightforward because we want to have only one instance of logger per class.

Also, `logger()` returns a delegate object, which means that the object will be created on its first access and will return the same value (object) on subsequent accesses.

There's more...

Anko is an Android library that uses Kotlin and makes Android development easier with the help of extension functions. It provides **Anko-logger**, which you can use if you don't want to write your own logger. It is included in *anko-commons*, which also has a lot of interesting things to make it worthwhile to include it in your Android projects that use Kotlin.

In Anko, a standard implementation of logger will look something like this:

```
class SomeActivity : Activity(), AnkoLogger {
    private fun someMethod() {
        info("London is the capital of Great Britain")
        debug(5) // .toString() method will be executed
        warn(null) // "null" will be printed
    }
}
```

As you can see, you just need to implement `AnkoLogger` and you are done.

Each method has two versions: plain and lazy (inlined):

```
info("String " + "concatenation")

info { "String " + "concatenation" }
```

The lambda result will be calculated only if `Log.isLoggable(tag, Log.INFO)` is true.

See also

To know more about delegated properties, refer to the **Working with delegated Properties** recipe in `Chapter 3`, *Classes and Objects*.

Escaping for Java identifiers that are keywords in Kotlin

Kotlin was designed with *interoperability* in mind. The existing code in Java can be called from Kotlin code smoothly, but since Java has different keywords than Kotlin, we sometimes run into issues when we call Java method with a name similar to Kotlin keyword. There is a workaround in Kotlin, which allows a method to be called having a name representing a Kotlin keyword.

Getting ready

Ensure that you have access to a code editor where you can write and run the code.

How to do it...

Create a Java class with a method name equal to any Kotlin keyword. I am using `is` as the method name, so my Java class looks as follows:

```
public class ASimpleJavaClass {
    static void is(){
        System.out.print("Nothing fancy here");
    }
}
```

Now try calling that method from Kotlin code. If you are using any code editor with the autocomplete feature, it automatically encloses the method name in backticks (` ` `):

```
fun main(args: Array<String>) {
    ASimpleJavaClass.`is`()
}
```

Similar is the case with other keywords in Kotlin that are qualified identifiers in Java.

How it works...

According to Kotlin's documentation, some of the Kotlin keywords are valid identifiers in Java: `in`, `object`, `is`, and so on. If a Java library uses a Kotlin keyword for a method, you can still call the method, escaping it with the backtick (` ` `) character.

The following are the keywords in Kotlin:

package	as	typealias	class	this	super	val
var	fun	for	null	true	false	is
in	throw	return	break	continue	object	if
try	else	while	do	when	interface	typeof

Disambiguating using the "as" keyword to locally rename the clashing entity

Disambiguation refers to the removal of ambiguity by making something clear. Importing a library or a class in code is a daily routine of a programmer. It's pretty easy to import files into the code in every language, thanks to the great code editors nowadays.

However, what happens if you try to import two classes into a file? Though you should always try to have different names for different classes, sometimes it's unavoidable. For example, in the case of different libraries having the same name for their classes. In Java, there is a workaround; you have to use the full qualifier, which looks something like this:

```
class X {
    com.very.very.long.prefix.bar.Foo a;
    org.other.very.very.long.prefix.baz.Foo b;
    ...
}
```

Dirty, isn't it? Now, let's see how Kotlin addresses it gracefully.

Getting ready

Ensure that you have a code editor on which you can write and run the code. To test things out, you can create two classes with the same name but under different packages. Refer to the example here:

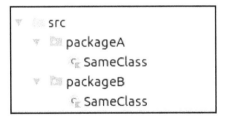

How to do it...

In the following steps and examples, we will see how we can disambiguate classes of the similar name using Kotlin's keyword.

1. In Kotlin, you can disambiguate using the `as` keyword to locally rename the clashing entity. So in Kotlin, it will look as shown:

```
import foo.Bar // Bar is accessible
import bar.Bar as bBar // bBar stands for 'bar.Bar'
```

2. Then, access their methods like this:

```
Bar.methodOfFooBar()
bBar.methodOfBarBar()
```

For example, let's see the use of the `as` keyword to disambiguate two classes having the same name (`SomeClass.kt`), but in different packages:

SameClass.kt (packageA)

```
package packageA
class SameClass {
  companion object {
    fun methodA(){
      println("Method a")
    }
  }
}
```

SameClass.kt (packageB)

```
package packageB
class SameClass {
  companion object {
    fun methodB(){
      println("Method b")
    }
  }
}
```

`HelloWorld.kt` is the class that uses classes with similar names:

```
import packageA.SameClass as anotherSameClass
import packageB.SameClass
fun main(args: Array<String>) {
    anotherSameClass.methodA()
    SameClass.methodB()

}
```

Doing bit manipulations in Kotlin

Kotlin provides several functions (in infix form) to perform bitwise and bit shift operations. In this section, we will learn to perform bit-level operation in Kotlin with the help of examples.

Bitwise and bit shift operators are used on only two integral types—Int and Long—to perform bit-level operations.

Getting ready

Here's the complete list of bitwise operations (available for Int and Long only):

- `shr(bits)`: signed shift right (Java's >>)
- `ushr(bits)`: unsigned shift right (Java's >>>)
- `and(bits)`: bitwise and
- `or(bits)`: bitwise or
- `xor(bits)`: bitwise xor
- `inv()`: bitwise inversion

How to do it...

Let's check out a few examples to understand the bitwise operations.

Or

The `or` function compares the corresponding bits of two values. If either of the two bits is 1, it gives 1, and it gives 0 if not.

Consider this example:

```
fun main(args: Array<String>) {
    val a=2
    val b=3
    print(a or b)
}
```

The following is the output:

```
3
```

Here's the explanation of the preceding example:

2 = 10 (Binary format)

3 = 11 (Binary format)

Bitwise OR of 2 and 3 that is

in binary

10 OR 11

11 = 3 (Decimal format)

and

The `and` function compares the corresponding bits of two values. If either of the two bits is 0, it gives 0, if not and both bits are 1, it gives 1.

Consider this example:

```
fun main(args: Array<String>) {
    val a=2
    val b=3
    print(a and b)
}
```

This is the output:

        2

Let's look at the explanation:

        2 = 10 (Binary format)

        3 = 11 (Binary format)

        Bitwise AND of 2 and 3

    in binary

        10 AND 11

        10 = 2 (Decimal format)

xor

The `xor` function compares the corresponding bits of two values. If the corresponding bits are the same, it gives 0, and if they are different, it gives 1.

Look at this example:

```
fun main(args: Array<String>) {
    val a=2
    val b=3
    print(a xor b)
}
```

Given is the output:

        1

Here's the explanation:

        2 = 10 (Binary format)

        3 = 11 (Binary format)

        Bitwise XOR of 2 and 3

in binary

10 XOR 11

01 = 1 (Decimal format)

inv

The `inv` function simply inverts the bit patterns. If the bit is 1, it makes it 0 and vice versa.

Here's an example:

```
fun main(args: Array<String>) {
    val a=2
    print(a.inv())}
```

This is the output:

```
-3
```

The following is the explanation:

2 = 10 (Binary format)

Bitwise complement of 2 = 01, but the compiler shows 2's complement of that number, which is the negative notation of the binary number.

2's complement of an integer n is equal to -(n+1).

shl

The `shl` function shifts the bit pattern to the left by the specified number of bits.

Consider this example:

```
fun main(args: Array<String>) {
    println( 5 shl 0)
    println( 5 shl 1)
    println( 5 shl 2)
}
```

This is the output:

```
5
10
20
```

Here's the explanation:

5 = 101 (Binary format)

101 Shift left by 0 bits = 101

101 Shift left by 1 bits = 1010 (10 in Decimal)

101 Shift left by 2 bits = 10100 (20 in Decimal)

shr

The shr function shifts the bit pattern to the right by the specified number of bits.

Take this example into consideration:

```
fun main(args: Array<String>) {
        println( 5 shr 0)
        println( 5 shr 1)
        println( 5 shr 2)
}
```

Given here is the output:

```
5
2
1
```

The following is the explanation:

5 = 101 (Binary format)

101 Shift right by 0 bits = 101

101 Shift right by 1 bits = 010 (2 in Decimal)

101 Shift right by 2 bits = 001 (1 in Decimal)

ushr

The `ushr` function shifts the bit pattern to the right by the specified number of bits, filling the leftmost with 0s.

Here's an example:

```
fun main(args: Array<String>) {
    println( 5 ushr 0)
    println( 5 ushr 1)
    println( 5 ushr 2)
}
```

This will output the following:

```
5
2
1
```

This is its explanation:

5 = 101 (Binary format)

101 Shift right by 0 bits = 101

101 Shift right by 1 bits = 010 (2 in Decimal)

101 Shift right by 2 bits = 001 (1 in Decimal)

How it works...

The bitwise operators in Kotlin aren't built-in operators like in Java, but they can still be used as an operator. Why? Look at its implementation:

```
public infix fun shr(bitCount: Int): Int
```

You can see that the method has the `infix` notation, which enables it to be called as an `infix` expression.

Parsing String to Long, Double, or Int

Kotlin makes it really easy to parse String into other data types, such as Long, Integer, or Double.

In JAVA, `Long.parseLong()`, or the `Long.valueOf()` static method is used, which parses the string argument as a signed decimal long and returns a long value, and similarly for other data types such as Int, Double, and Boolean. Let's see how to achieve it in Kotlin.

Getting ready

You just need a Kotlin editor to write and run your code. We'll use conversion of Long as an example to discuss parsing with string. Conversion to other data types is quite similar.

How to do it...

To parse the string to a Long data type, we use the `.toLong()` method with the string. It parses the string as a Long number and returns the result. It throws `NumberFormatException` if the string is not a valid representation of a number. Later, we will see examples for this.

Converting String to Long

Here's an example that shows parsing of string to Long:

```
fun main(args: Array<String>) {
  val str="123"
  print(str.toLong())
}
```

When you run the preceding code, you will see this output:

```
123
```

If you don't want to deal with the exceptions, you can use `.toLongOrNull()`. This method parses the string as a Long and returns the result, or null if the string is not a valid representation of a number.

Converting string to Long using string.toLongOrNull()

In this example, we will see how we can parse a string using the `.toLongOrNull()` method:

```
fun main(args: Array<String>) {
  val str="123.4"
  val str2="123"
  println(str.toLongOrNull())
  println(str2.toLongOrNull())
}
```

On running the preceding program, the following output is generated:

```
null 123
```

Converting with special radix

All the preceding examples use the base (radix) 10. There are cases when we wish to convert a String to Long but using another base. Both `string.toLong()` and `string.toLongOrNull()` can receive a custom radix to be used in the conversion. Let's take a look at its implementation:

- `string.toLong(radix)`:
  - This parses the string as a `[Long]` number and returns the result
  - `@throws NumberFormatException` if the string is not a valid representation of a number
  - `@throws IllegalArgumentException` when `[radix]` is not a valid radix for string to number conversion
- `string.toLongOrNull(radix)`:
  - This parses the string as a `[Long]` number and returns the result or null if the string is not a valid representation of a number
  - `@throws IllegalArgumentException` when `[radix]` is not a valid radix for string to number conversion

Parsing string to Long with special radix

In the preceding examples, we were parsing strings with radix 10, that is, decimals. By default, the radix is taken as 10, but there are certain situations where we need different radix. For example, in case of parsing a string into a binary or octal number. So now, we will see how to work with radix other than the decimal. Though you can use any valid radix, we will show examples that are most commonly used, such as binary and octal.

- **Binary**: Since a binary number is made from 0 and 1, the radix used is 2:

```
fun main(args: Array<String>) {
       val str="11111111"
       print(str.toLongOrNull(2))    }
```

On running the preceding program, the following output is generated:

```
255
```

- **Octal**: The octal numeral system, or oct for short, is the base-8 number system and uses the digits 0 to 7. Hence, we will use 8 as a radix:

```
fun main(args: Array<String>) {
       val str="377"
         print(str.toLongOrNull(8))
   }
```

On running the preceding program, this output is generated:

```
255
```

- **Decimal**: The decimal system has 10 numbers in it (0-9); hence, we will use 10 as radix. Note that radix as 10 is used by default in the methods without the radix arguments (`.toLong()` , `.toLongOrNull()`):

```
fun main(args: Array<String>) {
       val str="255"
         print(str.toLongOrNull(10))
   }
```

On running the preceding program, the following output is generated:

```
255
```

How it works...

Kotlin uses String's extension functions such as `.toLong()` and `toLongOrNull()` to make things easier. Let's dive into their implementation.

- For `Long`, use this:

```
public inline fun String.toLong(): Long =
java.lang.Long.parseLong(this)
```

As you can see, internally, it also calls the `Long.parseLong(string)` Java static method, and it is similar to the other data types.

- For `Short`, it's the following:

```
public inline fun String.toShort(): Short =
java.lang.Short.parseShort(this)
```

- Use this for `Int`:

```
public inline fun String.toInt(): Int =
java.lang.Integer.parseInt(this)
```

- For parsing with Radix, use the following:

```
public inline fun String.toLong(radix: Int): Long =
java.lang.Long.parseLong(this, checkRadix(radix))
```

The `checkRadix` method checks whether the given `[radix]` is valid radix for string to number and number to string conversion.

There's more...

Let's quickly see a few other extension functions provided by Kotlin to parse String:

- `toBoolean()`: Returns `true` if the content of this string is equal to the word *true*, ignoring case, and `false` otherwise.
- `toShort()`: Parses the string as a `[Short]` number and returns the result. Also, it throws `NumberFormatException` if the string is not a valid representation of a number.

- `toShort(radix)`: Parses the string as a `[Short]` number and returns the result, throws `NumberFormatException` if the string is not a valid representation of a number, and throws `IllegalArgumentException` when `[radix]` is not a valid radix for the string to number conversion.
- `toInt()`: Parses the string as an `[Int]` number and returns the result and throws `NumberFormatException` if the string is not a valid representation of a number.
- `toIntOrNull()`: Parses the string as an `[Int]` number and returns the result or `null` if the string is not a valid representation of a number.
- `toIntOrNull(radix)`: Parses the string as an `[Int]` number and returns the result or `null` if the string is not a valid representation of a number, or `@throws IllegalArgumentException` when `[radix]` is not a valid radix for string to number conversion.
- `toFloat()`: Parses the string as a `[Float]` number and returns the result, and `@throws NumberFormatException` if the string is not a valid representation of a number.
- `toDouble()` : Parses the string as a `[Double]` number and returns the result, and `@throws NumberFormatException` if the string is not a valid representation of a number.

Using String templates in Kotlin

Kotlin packs great features with commonly used data type String. One of the really cool features is String templates. This feature allows Strings to contain template expression.

In Java, you had to use **StrSubstitutor** (`https://commons.apache.org/proper/commons-text/javadocs/api-release/org/apache/commons/text/StrSubstitutor.html`) and a map to go with it. A template expression in Java will look as follows:

```
Map<String, String> valuesMap = new HashMap<String, String>();
valuesMap.put("city", "Paris");
valuesMap.put("monument", "Eiffel Tower");
String templateString ="Enjoyed ${monument} in ${city}.";
StrSubstitutorsub=newStrSubstitutor(valuesMap);
String resolvedString =sub.replace(templateString);
```

Kotlin eases out the pain in writing template expressions and makes it fun, concise, and a lot less verbose.

Using String templates, you can embed a variable or expression inside a string without string concatenation. So, let's get started!

How to do it...

In the following steps, we will learn how to use String templates:

1. In Kotlin, the template expression starts with a $ sign.
2. The syntax of string templates is as follows:

```
$variableName
```

Alternatively, it is this:

```
${expression}
```

3. Let's check out a few examples:

- Consider the example of a String template with variable:

```
fun main(args: Array<String>) {
    val foo = 5;
    val myString = "foo = $foo"
    println(myString)
}
```

The output of the preceding code will be `foo = 5`.

- Consider the example of a String template with expression:

```
fun main(arr: Array<String>){
  val lang = "Kotlin"
  val str = "The word Kotlin has ${lang.length} characters."
  println(str)
}
```

- Consider the example of a String template with raw string:
    - **Raw string**: A string consisting of newlines without writing \n and arbitrary string. It's a raw string and is placed in triple quotes ("""):

```
fun main(args: Array<String>) {
    val a = 5
    val b = 6
```

```
            val myString = """
            ${if (a > b) a else b}
        """
            println("Bigger number is: ${myString.trimMargin()}")
        }
```

When you run the program, the output will be `Bigger number is: 6`.

How it works...

The use of String template with a variable name is quite straightforward. Earlier, we used to concatenate the strings, but now we can just specify the variable with the `$` symbol before it.

When the string template is used as an expression, the expression inside the `${..}` is evaluated first and the value is concatenated with the string. In the preceding example (String template with raw string), the `${if (a > b) a else b}` expression is evaluated and its value, that is 6, is printed with the string.

There's more...

String templates also come in handy with String properties and functions. Here's an example:

```
fun main(args: Array<String>) {
    val str1="abcdefghijklmnopqrs"
    val str2="tuvwxyz"
    println("str1 equals str2 ? = ${str1.equals(str2)}")
    println("subsequence is ${str1.subSequence(1,4)}")
    println("2nd character is ${str1.get(1)}")
}
```

Here's the output:

```
str1 equals str2 ? = false
subsequence is bcd
2nd character is b
```

2
Control Flow

The following recipes will be covered in this chapter:

- Assigning result to an expression using the `if` keyword

- Using range with the `when` expression

- Using `when` with custom objects

- Using `try-catch` as an expression

- How to write a swap function in Kotlin using the `also` function

- How to throw a custom exception in Kotlin

- How to make a multiconditional loop in Kotlin

Introduction

Control flows are the basic building block of every programming language. What's different in Kotlin is that you can use a few of those control flows as an expression, such as `try-catch`, `if-else`, `when`, and so on. In this chapter, we will go through some of the control flows offered by Kotlin and learn to use them. Furthermore, we will also see how they provide much more power than Java control flows. So let's get started!

Assigning result to an expression using the if keyword

In Kotlin `if` is special because it returns values. That is why we can use an `if` statement to assign values to a result. This removes the need for a ternary operator in Kotlin. Let's see how we can use `if` statements to assign value.

Getting ready

You need to install a preferred development environment that compiles and runs Kotlin. You can also use the command line for this purpose, for which you need Kotlin compiler installed, along with JDK. I am using the command line for compiling and running my Kotlin code for this recipe.

How to do it...

Create a file and name it `ifWithKotlin.kt`. You can name it anything; it need not be the same as the class name because it is in Java. Now, to get started, you should always declare the main method because the Java virtual machine starts execution by invoking the `main` method of the specified class.

The `main` method is as follows:

```
fun main(args: Array<String>) { }
```

1. Let's try a basic `if` statement in a traditional way to understand how it works:

```
fun main(args: Array<String>) {
    var x:Int
    if(10>20){
        x = 5
    }
    else{
        x = 10
    }
    println("$x")
}
```

In this code block, we assign a value to x in the `if` and `else` block and then print it.

Chapter 2

2. Now, let's try the same thing the Kotlin way:

```
fun main(args: Array<String>) {
    var x:Int = if(10>20)   5   else   10
    println("$x")
}
```

In this code block, we assigned a value returned by the `if-else` block to `x`. Note how we've used an `if` statement as a part of the expression on the right-hand side of the expression.

3. Let's see what else can we do. In the following example, we will try to return something from the expression using the `if` statement:

```
fun main(args: Array<String>) {
    var x:Int
    x = if(10>20) {
            doSomething()
            25
    }
    else if (12<13) {
        26
    }
    else{
        27
    }
    println("$x")
}
fun doSomething() {
    var a = 6
    println("$a")
}
```

Note how we used the whole block of `if-else`. In this case, the `if` block returns the last statement of the block.

4. Finally, let's try a more complicated example using a nested `if-else`. This will help us understand how values are returned in a nested `if-else` structure:

```
fun main(args: Array<String>) {
    var x:Int
    x = if(10<20) {
        if(4 == 3){
            56
        }
        else{
```

[55]

```
                96
            }
        }
        else if (12>13) {
            26
        }
        else{
            27
        }
        println("$x")
    }
```

//Output: 96

So, if we nest an if–else block and if the last statement of that if–else block is another if–else statement, the value returned by the nested if–else is returned by the enclosing one. As you can see, 96 is returned by the else block inside the if(10<20) block.

5. What happens if the if–else block is not the last statement, like in this example:

```
fun main(args: Array<String>) {
var x:Int
x = if(10<20)  {
        if(4 == 3){
                56
        }
        else{
                96
        }
        565
    }
    else if (12>13) {
        26
    }
    else{
        27
    }
    println("$x")
}
```

Clearly, the value returned by the nested if—else is not being used, and the Kotlin compiler also warns us of this. The reason behind this is because the if—else block is not the last statement of the parent if—else block, which is why the returned value is not being used.

```
○ ● ◎  /bin/bash
                          /bin/bash 80x24
rashi Desktop $ kotlinc ifWithKotlin.kt -include-runtime -d ye.jar
ifWithKotlin.kt:7:13: warning: the expression is unused
            56
            ^

ifWithKotlin.kt:10:13: warning: the expression is unused
            96
            ^

rashi Desktop $ java -jar ye.jar
6
565
rashi Desktop $ ▉
```

Try playing around with the values and logic to see what else you can do with if-else.

The key thing to always remember is that the last statement of the if-else block is returned, which is why it can be used to assign values to any variable.

There's more...

We have used string templates in print statements. Note how we are able to access a variable using the $ symbol before the name of a variable:

```
println("$a is a number something")
```

We can also put a piece of code in strings that are evaluated and whose results are concatenated into the string. In this case, $ is followed by { }, inside which we put our code:

```
println("some variable whose value: ${if(a < 100) 25 else 29}")
```

Using range with the when expression

In Kotlin, when is like a super-powered switch control statement. However, that's not all it can do. There's a lot of amazing logic that you can build with the when statement, one example of which is using a range with the when statement. We will take a look at that in this recipe.

Getting ready

You need to install a preferred development environment that compiles and runs Kotlin. You can also use the command line for this purpose, for which you need Kotlin compiler installed, along with JDK. I am using the command line to compile and run my Kotlin code for this recipe.

How to do it...

First, let's create a file, name it whenWithRanges.kt, and follow these steps:

1. Let's try a basic when statement to understand how it works:

```
fun main(args: Array<String>) {
    val x = 12
    when(x){
        12 -> println("x is equal to 12")
        4 -> println("x is equal to 4")
        else -> println ("no conditions match!")
    }
}
```

So basically, this code block works like a switch case statement, and it can also be implemented using an if statement.

2. Now, let's see if x lies between 1 to 10 or outside it:

```
fun main(args: Array<String>) {
    val x = 12
    when(x){
        in (1..10) -> println("x lies between 1 to 10")
        !in (1..10) -> println("x does not lie between 1 to 10")
    }
}
```

3. Let's see what else we can do. In the following example, we will work with different types of conditions that can be used inside a when statement:

```kotlin
fun main(args: Array<String>) {
    val x = 10
    when(x){
        magicNum(x) -> println("x is a magic number")
        in (1..10) -> {
            println("lies between 1 to 10, value: ${if(x < 20) x else 0}")
        }
        20,21 -> println("$x is special and has direct exit access")
        else -> println("$x needs to be executed")
    }
}
fun magicNum(a: Int): Int {
 return if(a in (15..25)) a else 0
 }
```

4. Finally, let's try a more complicated example of using data classes. In this example, we will see how to use when with objects:

```kotlin
fun main(args: Array<String>) {
    val x = ob(2, true, 500)
    when(x.value){
        magicNum(x.value) -> println("$x is a magic number and ${if(x.valid) "valid" else "invalid"}")
        in (1..10) -> {
            println("lies between 1 to 10, value: ${if(x.value < x.max) x.value else x.max}")
        }
        20,21 -> println("$x is special and has direct exit access")
        else -> println("$x needs to be executed")
    }
 }
 data class ob(val value: Int, val valid: Boolean, val max: Int)
 fun magicNum(a: Int): Int {
 return if(a in (15..25)) a else 0
 }
```

Here's how it looks after compiling and running the program:

```
/bin/bash
                          /bin/bash 80x24
rashi Desktop $ kotlinc whenWithRanges.kt -include-runtime -d ye.jar
rashi Desktop $ java -jar ye.jar
lies between 1 to 10, value: 2
rashi Desktop $
```

Try playing around with the values and logic to see what else you can do with such a small block of code in Kotlin using when.

How it works...

In the preceding examples, the first example is the most basic when statement; we are directly comparing *x*'s value to 12 and 4, and if no conditions match, we are simply executing the else statement. It is like an if else if else statement.

In the second example, we check whether x lies between 1 to 10 in the first statement inside the when block, and in the second statement, we check whether x does not lie between 1 to 10. That's how we work with ranges in when. Basically, in when, we can check whether x lies in a range or exists in a collection using the in keyword. The syntax is as follows:

```
when(x) {
    In collection_or_range -> // do something
}
```

In the third example, we use a function to check whether x equals the value of the expression magicNum(x). So we can also use expressions and functions in place of constants and ranges to compare x.

In the fourth example, we explore the power of the when statement using a data class instead of a primitive data type in when. Note how we are able to access all properties of x inside when and also play with them.

There's more...

We have already seen how we can use string templates with expressions in print statements. Remember how we were able to access a variable using the $ symbol before the name of a variable:

```
println("$x is a magic number")
```

We can also put a piece of code in a string, which is then evaluated and whose results are concatenated into the string. In this case, $ is followed by { }, inside which we put our code:

```
println("lies between 1 to 10, value: ${if(x.value < x.max) x.value
else x.max}")
```

Using when with custom objects

In Kotlin, when is already so powerful but did you know you can also use custom objects in when? Amazing, right? Let's go about implementing it.

Getting ready

You need to install a preferred development environment that compiles and runs Kotlin. You can also use the command line for this purpose, for which you need Kotlin compiler installed, along with JDK. I am using the command line to compile and run my Kotlin code for this recipe.

How to do it...

Create a file and name it whenWithObject.kt, and then, let's try when with a custom object. In this example, we will create an object with some properties and try to match it in a when statement:

```
fun main(args: Array<String>) {
    val x = ob(2, true, 500)
    when(x){
        ob(2, true, 500) -> println("equals correct object")
        ob(12, false, 800) -> {
            println("equals wrong object")
        }
        else -> println("does not match any object")
```

```
            }
        }
        data class ob(val value: Int, val valid: Boolean, val max: Int)
```

Here's the output of the preceding code block:

If you try to compare a different object type in `when`, it throws an error `error: incompatible types` because we are trying to compare objects of different types.

How it works...

In Kotlin, `when` basically works with equality in the background, so we can compare objects as long as their types are the same.

Using try–catch as an expression

Exceptions in Kotlin are both similar and different compared to those in Java. In Kotlin, `Throwable` is the superclass of all the exceptions, and every exception has a stack trace, message, and an optional cause.

The structure of `try–catch` is also similar to that used in Java. In Kotlin, here's how a `try–catch` statement looks:

```
try {
// some code to execute
}
catch (e: SomeException) {
// exception handler
}
```

```
finally {
// optional finally block
}
```

At least one `catch` block is mandatory and the `finally` block is optional, and so it can be omitted.

In Kotlin, `try-catch` is special as it enables it to be used as an expression. In this article, we will see how we can use `try-catch` as an expression.

Getting ready

You need to install a preferred development environment that compiles and runs Kotlin. You can also use the command line for this purpose, for which you need Kotlin compiler installed, along with JDK. I am using IntelliJ IDE to compile and run my Kotlin code for this recipe.

How to do it...

Let's write a simple program that takes in a number as an input and assigns its value to a variable. If the entered value is not a number, we catch the `NumberFormatException` exception and assign −1 to that variable:

```
fun main(args: Array<String>) {
 val str="23"
 val a: Int? = try { str.toInt() } catch (e: NumberFormatException)
{ -1 }
 println(a)
 }
```

This is the output:

```
Output: 23
```

Now, let's try something crazy and deliberately try to throw the exception:

```
fun main(args: Array<String>) {
 val str="abc"
 val a: Int? = try { str.toInt() } catch (e: NumberFormatException)
{ -1 }
 println(a)
 }
```

This is the output:

```
Output: -1
```

The usage of try-catch will help you a lot in edge cases as they can be used as an expression.

How it works...

The reason we can use try-catch as an expression is that both try and throw are expressions in Kotlin and hence can be assigned to a variable.

When you use try-catch as an expression, the last line of the try or catch block is returned. That's why, in the first example, we got 23 as the returned value and we got -1 in the second example.

Here, one thing to note is that the same thing doesn't apply to the finally block—that is, writing the finally block will not affect the result:

```
fun main(args: Array<String>) {
    val str="abc"
    val a:Int = try {
            str.toInt()
        } catch (e: NumberFormatException) {
            -1
        } finally {
            -2
        }
    println(a)
}

Output: -1
```

As you can see, writing the finally block doesn't change anything.

There's more...

In Kotlin, all the exceptions are unchecked, which means that we don't need to apply try-catch at all. This is quite different than Java, where if a method throws an exception, we need to surround it with try-catch.

Here's an example of an IO operation in Kotlin:

```
fun fileToString(file: File) : String {
//readAllBytes throws IOException, but we can omit catching it
fileContent = Files.readAllBytes(file)
return String(fileContent)
}
```

As you can see, we don't need to wrap things with `try-catch` if we don't want to. In Java, we couldn't proceed without handling this exception.

How to write a swap function in Kotlin using the also function

Swapping two numbers is one of the most common things you do in programming. Most of the approaches are quite similar in nature: Either you do it using a third variable or using pointers.

In Java, we don't have pointers, so mostly we rely on a third variable.

You can obviously use something as mentioned here, which is just the Kotlin version of Java code:

```
var a = 1
var b = 2
run { val temp = a; a = b; b = temp }
println(a) // print 2
println(b) // print 1
```

However, in Kotlin, there is a very quick and intuitive way of doing it. Let's see how!

Getting ready

You need to install a preferred development environment that compiles and runs Kotlin. You can also use the command line for this purpose, for which you need Kotlin compiler installed, along with JDK. I am using IntelliJ IDE to compile and run my Kotlin code for this recipe.

How to do it...

In Kotlin, we have a special function, `also`, that we can use to swap two numbers. Here's the code to go with it:

```
var a = 1
var b = 2
a = b.also { b = a }
println(a) // print 2
println(b) // print 1
```

We were able to achieve the same thing without using any third variable.

How it works...

To understand the preceding example, we need to understand the `also` function in Kotlin. The `also` function takes the receiver, performs some operation, and returns the receiver. In simple words, it passes an object and returns the same object.

Applying the `also` function on an object is like saying "do this as well" to that object.

So, we called the `also` function on `b`, did an operation (assigning the value of `a` to `b`), and then returned the same receiver that we got as an argument:

```
var a = 1
 var b = 2
a = b.also {
        b = a   // p
        println("it=$it : b=$b : a=$a") // prints it=2:b=1:a=1
     }
println(a) // print 2
println(b) // print 1
```

There's more...

The `apply` function is quite similar to the `also` function, but they have a subtle difference. To understand that, let's look at their implementation first:

- The `also` function:

```
public inline fun <T> T.also(block: (T) -> Unit): T { block(this);
return this }
```

- The `apply` function:

```
public inline fun <T> T.apply(block: T.() -> Unit): T { block();
return this }
```

In `also`, the block is defined as `(T) -> Unit`, but it is defined as `T.() -> Unit` in `apply()`, which means there is an implicit `this` inside the `apply` block. However, to reference it in `also`, we need `it`.

So a code using `also` will look like this:

```
val result = Dog(12).also { it.age = 13 }
```

The same will look like this using `apply`:

```
val result2 =Dog(12).apply {age = 13 }
```

The age of the resulting object will be the same in both the cases, that is, `13`.

How to throw a custom exception in Kotlin

Sometimes, there are cases where you want to create your own exception. If you are creating your own exception, it's known as a **custom exception** or **user-defined exception**.

These are used to customize the exception according to a specific need, and using this, you can have your own exception and a message. In this recipe, we will see how to create and throw a custom exception in Kotlin.

Getting ready

You need to install a preferred development environment that compiles and runs Kotlin. You can also use the command line for this purpose, for which you need Kotlin compiler installed, along with JDK. I am using IntelliJ IDE to compile and run my Kotlin code for this recipe.

How to do it...

All the exceptions have `Exception` as their superclass, so we need to extend that class.

Here's what our custom exception looks like:

```
class CustomException(message:String): Exception(message)
```

Since the `Exception` superclass has a constructor that can take in a message, we've passed it with the help of the constructor of `CustomException`.

Now, if you have to `throw` an `Exception`, you will need to simply do the following:

```
throw CustomException("Threw custom exception")
```

The output will be something like this:

```
Run    HelloWorldKt
    /usr/lib/jvm/java-1.8.0-openjdk-amd64/bin/java ...
    Exception in thread "main" CustomException: Threw custom exception
        at HelloWorldKt.main(HelloWorld.kt:4)

    Process finished with exit code 1
```

How it works...

Let's take a look at the implementation of the `Exception` class:

```
public class Exception extends Throwable {
 static final long serialVersionUID = -3387516993124229948L;
public Exception() {
 }
public Exception(String var1) {
     super(var1);
 }.....
```

As you can see, we have a second constructor that takes a `String` as a parameter. In our `CustomException` class, we have supplied it by passing its message to the superclass's constructor. Also, you can create a custom exception with an empty constructor because `Exception` also has an empty constructor.

How to make a multiconditional loop in Kotlin

Conditional loops are common to any programming language you pick. If you apply multiple conditions on a loop, it is called a **multiconditional loop**. A simple example of a multiconditional loop in Java is illustrated here:

```
int[] data = {5,6,7,1,3,4,5,7,12,13};
for(int i=0;i<10&&i<data[i];i++){
    System.out.println(data[i]);
}
```

The preceding code on execution will print out 5, 6, and 7. Let's see how we can use a multiconditional loop in Kotlin. We will be looking at a functional approach to the same thing in Kotlin.

Getting ready

You need to install a preferred development environment that compiles and runs Kotlin. You can also use the command line for this purpose, for which you need Kotlin compiler installed, along with JDK. I am using IntelliJ IDE to compile and run my Kotlin code for this recipe.

How to do it...

The preceding multiconditional loop can be written in Kotlin like so:

```
(0..9).asSequence().takeWhile {
    it<numbers[it]
}.forEach
    println("$it - ${data[it]}")
}
```

It's nice, clean, and definitely not an eyesore.

How it works...

We used `takeWhile`, which returns a sequence containing first elements satisfying the given predicate (in this case, `i<data[i]`).

Though `takeWhile` returns the first elements that satisfy the given predicate, you might be tempted to think that it will first evaluate the complete range and then pass to `forEach`. That would have been the case if we hadn't used `.asSequence()`. We converted the range to a `Sequence<T>`, and because of this, it was lazily evaluated. In short, it won't process the whole set of items with `.takeWhile { ... }` and will only check them one by one when `.forEach { ... }` is up to process the next item.

Let's try to understand this with the help of an example. First, we will work with an eager evaluation over `Iterable<T>`.

This is the eager version, which evaluates the first function before moving on to the next one:

```
(0..9).takeWhile {
    println("Inside takeWhile")
    it<numbers[it]
}.forEach {
    println("Inside forEach")
}
```

This is the output:

```
Inside takeWhile
Inside takeWhile
Inside takeWhile
Inside takeWhile
Inside forEach
Inside forEach
Inside forEach
```

As you can see, the range was first processed with `takeWhile` (which returned 0, 1, 2) and was then sent for processing to `forEach`.

Now, let's see the lazy version:

```
(0..9).asSequence().takeWhile {
    println("Inside takeWhile")
    it<numbers[it]
}.forEach {
    println("Inside forEach")
}
```

Here's the output:

```
Inside takeWhile
  Inside forEach
  Inside takeWhile
  Inside forEach
  Inside takeWhile
  Inside forEach
  Inside takeWhile
```

As you can see in the preceding example, `takeWhile` is evaluated only when `forEach` is used to processes an item. This is the nature of `Sequence<T>`, which performs lazily where possible.

3
Classes and Objects

The following recipes will be covered in this chapter:

- Initializing body of constructor

- Converting one data type into another

- How to type check an object

- How to work with an abstract class in Kotlin

- How to iterate over a class's properties in Kotlin

- How to work with inline properties

- How to work with nested class

- Getting class in Kotlin

- Working with delegated properties

- Working with enums

Introduction

In this chapter, you will be introduced to recipes related to object-oriented programming in Kotlin. Using an OOP approach, you can divide complex problems into smaller problems by creating objects. There are a few differences in Kotlin's style of OOP as compared to Java—for example, in Kotlin, all the classes are closed (final) by default, and if you want them to be extensible, you need to make them open by using an open keyword. Not only for classes—even the methods are final by default, and you need an open keyword for them as well. With Kotlin much less code is needed to work with classes and objects. Oh! By the way, did I tell you that we don't even need to use the new keyword while creating the object? So, creating a new object in Kotlin is as simple as this:

```
var person=Person()
```

The preceding code will create a mutable object of type Person, because we have used var as a modifier. A mutable object means that it can change its value. If you want to create an immutable object, you do it using the val keyword. So the same example will look as follows:

```
val person=Person()
```

So, let's begin looking at some recipes that will help you with object-oriented programming in Kotlin.

Initializing body of constructor

In the Java world, we used to initialize fields of the class in the constructor, as shown in this code:

```
class Student{
 int roll_number;
 String name;
 Student(int roll_number,String name){
    this.roll_number =roll_number;
    this.name = name;
 }
}
```

So, if the argument's name was similar to that of the property (which was usually the case for making the code more readable), we needed to use this keyword. In this recipe, we will see how to implement the same thing in Kotlin (obviously with much less code).

Getting ready

You need an IDE to write and execute your code. I'll be using IntelliJ IDEA. We will create a Student class with name and roll_number as properties.

How to do it...

Let's look at the mentioned steps to initialize a constructor:

1. Kotlin provides a syntax that can initialize the properties with much less code. Here's what class initialization looks like in Kotlin:

```
class Student(var roll_number:Int, var name:String)
```

2. You don't even need to define the body of the class, and the initialization of properties takes place in the primary constructor only (the primary constructor is part of the class header). Obviously, you can either choose var or val, based on whether you need to keep your properties mutable or not. Now, if you try to create an object, you can do so with this:

```
var student_A=Student(1,"Rashi Karanpuria")
```

3. Just to confirm, let's try to print its properties to see whether we were able to initialize it or not:

```
println("Roll number: ${student_A.roll_number} Name:
${student_A.name}")
```

Here's the output:

```
Roll number: 1 Name: Rashi Karanpuria
```

4. However, if you want, you can also put default values in the constructor:

```
class Student constructor(var roll_number:Int, var
name:String="Sheldon")
```

5. Then, you can create objects such as this:

```
var student_sheldon= Student(25)      // Object with name Sheldon and
age 25

var student_amy=Student(25, "Amy")      // Object with name Amy and
age 25
```

6. If the class has a primary constructor, each secondary constructor needs to be delegated to the primary constructor, either directly or indirectly through another secondary constructor(s).

7. We use this keyword to delegate to another constructor of the same class:

```
class Person(val name: String) {
    constructor(name: String, lastName: String) : this(name) {
        // Do something maybe
    }
}
```

8. We can also have a situation where we have to initialize other things in the class, not necessarily just the class's properties. That situation could be opening database connections, for example. In Java, that was done in the constructor itself, but in Kotlin, we have an `init` block. The initialization code can be put into an `init` block:

```
class Student(var roll_number:Int,var name: String) {
    init {
        logger.info("Student initialized")
    }
}
```

9. Sometimes, we also initialize properties of a class by dependency injection. If you've worked with Dagger2, you must be familiar with objects being directly injected into the constructors of a class. To do so, we append the `@Inject` annotation before the constructor keyword. Whenever a constructor has an annotation or visibility modifier, we need to have the `constructor` keyword. An example of the constructor keyword is given below:

```
class Student @Inject constructor(compositeDisposable:
CompositeDisposable) { ... }
```

10. Here, we are injecting an object of the `CompositeDisposable` type into the constructor and, since we are using an annotation (`@Inject`) to do so, we need to apply the constructor keyword.

11. When you extend a class, you need to initialize the superclass. This is also very simple in Kotlin. If your class has a primary constructor, the base type must be initialized there, using the parameters of the primary constructor. Here's an example of the same:

```
class Student constructor(var roll_number:Int, var
name:String):Person(name)
```

12. However, sometimes a class may not have the primary constructor. In that case, each secondary constructor has to initialize the base type using the `super` keyword or can delegate to another constructor that does that. Also, different secondary constructors can call different constructors of the base type:

```
class Student: Person {
 constructor(name: String) : super(name)
 constructor(name: String, roll_number: Inte) :super(name)
 }
```

Converting one data type into another

In Java, we used to typecast by appending the desired type in front of variables like this:

```
String a = Integer.toString(10)
```

Also, in Java, numeric is directly converted to larger numeric types, but in Kotlin, this feature is not there for type safety—so how can we change one type of object to another in Kotlin? We will see it in this recipe.

Getting ready

You need to install a preferred development environment that compiles and runs Kotlin. You can also use the command line for this purpose, for which you need Kotlin compiler installed, along with JDK. I am using the online IDE available at `https://try.kotlinlang.org/` to compile and run my Kotlin code for this recipe.

How to do it...

Let's understand how to convert one data type into another by following the steps below:

1. Let's try a very basic example—trying to convert an `Int` to `Long` and `Float`:

```
fun main(args: Array<String>) {
    var a = 1
    var b: Float = a.toFloat()
    var c = a.toLong()
    println("$a is Int while $b is Float and $c is Long")
}
```

2. Similarly, `Long` can be converted to `Float` and `Int`, like this:

```
fun main(args: Array<String>) {
    var a = 1000000000000000000L
    var b: Float = a.toFloat()
    var c = a.toInt()
    println("$a is Long while $b is Float and $c is Integer")
}
```

The output of this code is as shown:

```
1000000000000000000 is Long while 9.9999998E17 is Float and
-1486618624 is Integer
```

3. Let's try a more interesting conversion with `Byte`, `Int`, and `Strings`:

```
fun main(args: Array<String>) {
    var a = 15623
    var b: Byte = a.toByte()
    var c = a.toString()
    println("$a is Int while $b is Byte and $c is String")
}
```

Here's a list of methods that can be used for type conversion in Kotlin:

- `toByte()`: Byte
- `toShort()`: Short
- `toInt()`: Int
- `toLong()`: Long
- `toFloat()`: Float
- `toDouble()`: Double
- `toChar()`: Char
- `toString()`: String

How it works...

Basically, Kotlin is a type-safe language and ensures that types cannot be directly converted in the language. Also, `String` is not the same as `String`? As expected, there is no method to convert a variable to a Boolean type. Conversion from a larger type to a smaller type is possible, but it might truncate the resulting values.

How to type check an object

One often needs to check if an object is of a particular type at runtime. With Java, we used an instance of a keyword; with Kotlin, it is the `is` keyword.

Getting ready

You need to install a preferred development environment that compiles and runs Kotlin. You can also use the command line for the purpose, for which you need Kotlin compiler installed along with JDK. I am using online IDE at `https://try.kotlinlang.org/` to compile and run my Kotlin code for this recipe.

How to do it...

Let's see how to type check an object in these steps:

1. Let's try a very basic example, trying `is` with string and integer. In this example, we will type check a string and an integer:

```
fun main(args: Array<String>) {
    var a : Any = 1
    var b : Any = "1"
    if (a is String) {
        println("a = $a is String")
    }
    else {
        println("a = $a is not String")
    }
    if (b is String) {
        println("b = $b is String")
    }
    else {
        println("b = $b is not String")
    }
}
```

2. Similarly, we can use `!is` to check whether the object is not of type `String`, like this:

```
fun main(args: Array<String>) {
    var b : Any = 1
    if (b !is String) {
```

```
            println("$b is not String")
        }
        else {
            println("$b is String")
        }
    }
```

If you remember how `when` works in Kotlin, we do not need to put in the `is` keyword, because Kotlin has a feature of smart cast and throws an error if the compared objects are not of the same type.

How it works...

Basically, the `is` operator is used to check the type of the object in Kotlin and `!is` is the negation of the `is` operator.

Kotlin compiler tracks immutable values and safe casts them wherever needed. This is how smart casts work; `is` is a safe cast operator, whereas an unsafe cast operator is the `as` operator.

There's more...

Let's try an example with the `as` operator, which is used for casting in Kotlin. It is an unsafe cast operator. The following code example throws `ClassCastException`, because we cannot convert an integer to string:

```
fun main(args: Array<String>) {
    var a : Any = 1
    var b = a as String
}
```

On the other hand, the following code runs successfully because of variable `a`, which, being of `Any` type, can be cast to `String`:

```
fun main(args: Array<String>) {
    var a : Any = "1"
    var b = a as String
    println(b.length)
}
```

How to work with an abstract class in Kotlin

Abstract classes are classes that cannot be instantiated, which means that we cannot create objects of an abstract class. The main inspiration behind using abstract classes is that we can inherit from them. When a class inherits from an abstract class, it implements all abstract methods of the parent class.

Getting ready

You need to install a preferred development environment that compiles and runs Kotlin. You can also use the command line for the purpose, for which you need Kotlin compiler installed along with JDK. I am using online IDE at `https://try.kotlinlang.org/` to compile and run my Kotlin code for this recipe.

How to do it...

Let's now see how to work with an `abstract` class in these steps:

1. The `abstract` keyword is used to declare an `abstract` class. Let's create an abstract class and try to inherit from it:

```
abstract class Mammal {
    abstract fun move(direction: String)
}
```

2. For a class to be a subclass of the `Mammal` class, we use the : operator, as in the following example. Pay attention to the `override` keyword used before the method implementation of the superclass:

```
class Dog : Mammal() {
    override fun move(direction: String) {
        println(direction)
    }
}
```

3. If we do not want a method to be implemented by the subclass, we do not declare it as `abstract` or `open`, as demonstrated in this example:

```
fun main(args: Array<String>) {
    var x = Dog()
    x.move("North")
    println(x.show(123))
}
class Dog : Mammal() {
    override fun move(direction: String) {
        println(direction)
    }
}
abstract class Mammal {
    fun show(y: Int): String {
        return y.toString()
    }
    abstract fun move(direction: String)
}
```

4. If we declare `init` blocks in each class, as follows, we get an output where superclass's `init` block is called first:

```
fun main(args: Array<String>) {
    var x = Dog()
    x.move("North")
    println(x.show(123))
}
class Dog : Mammal() {
    init {
        println ("Hey from Dog")
    }
    override fun move(direction: String) {
        println(direction)
    }
}
abstract class Mammal {
    init {
        println ("Hey from Mammal")
    }
    fun show(y: Int): String {
        return y.toString()
    }
    abstract fun move(direction: String)
}
```

The output of the final program is this:

```
Hey from Mammal
Hey from Dog
North
123
```

How it works...

The `Dog` class is a subclass of `Mammal` and inherits all its methods. The methods declared `abstract` are supposed to be implemented by the `Dog` class. The `show()` method is in `Mammal` but can be called by the `Dog` object, because the object created is of the `Mammal` type.

The `init` block of superclass is called before subclass.

How to iterate over a class's properties in Kotlin

Reflections in Kotlin allows us introspection of the structure of our program at runtime. This also enables us to introspect the class modifiers, methods, and properties.
In this recipe, we will see how we can iterate over the properties of a Kotlin class. So let's get started!

Getting ready

We'll be using IntelliJ IDEA IDE for coding purposes. We will create a `Student` class, which will have the `roll_number` and `name` properties. We will then see how we can iterate over its properties.

If you are not using IntelliJ IDE or Android Studio, you might need to include reflection library in your classpath. Head on over to `https://kotlinlang.org/docs/reference/reflection.html` to learn more about this.

How to do it...

In the following steps, we will see how to iterate over a class's properties:

1. Here's our `Student` class with the `roll_number` and `full_name` attributes:

```
class Student constructor(var roll_number:Int, var
full_name:String)
```

2. Now, we will be using a `for` statement, because we want to iterate over multiple properties that a class can have:

```
fun main(args: Array<String>) {
    var student=Student(2013001,"Aanand Shekhar Roy")
    for (property in Student::class.memberProperties) {
        println("${property.name} = ${property.get(student)}")
    }
}
```

This is the output:

```
full_name = Aanand Shekhar Roy
roll_number = 2013001
```

How it works...

The implementation is quite straightforward. We are able to achieve the introspection into the class's properties because we are using reflections and `memberProperties` is just one of the many functions of `KClass`.

One thing to note is that `memberProperties` returns all the non-extension properties declared in this class and all of its superclasses. Consider that we have a `Person` class, as follows:

```
open class Person{
    val isHuman:Boolean=true
}
```

Also, we extend our `Student` class with the `Person` class, and then the same code used earlier with the `memberProperties` method will result in an output as shown:

```
full_name = Aanand Shekhar Roy
roll_number = 2013001
isHuman = true
```

So, if you want to just iterate over the declared fields in the `Student` class, you will need the `declaredMemberProperties` method. Here's an example with `declaredMemberProperties`:

```
for (property in Student::class.declaredMemberProperties) {
    println("${property.name} = ${property.get(student)}")
}
```

This is the output:

```
full_name = Aanand Shekhar Roy
roll_number = 2013001
```

The preceding examples were for Kotlin `KClass`. Suppose you want to iterate over properties for a `Java Class<T>`—you can use a Kotlin extension property to get the Kotlin `KClass<T>`, from which you can proceed, for example, `something.javaClass.kotlin.memberProperties`.

There's more...

Check out the list (`https://kotlinlang.org/api/latest/jvm/stdlib/kotlin.reflect/-k-class/index.html`) of methods provided by Kotlin's Reflection library, with the help of which you can perform a lot of introspection at runtime.

How to work with inline properties

A great thing about Kotlin is high-order functions that let us use functions as parameters to other functions. However, they are objects, so they present memory overhead (because every instance is allocated space in heap, and we need methods for calling the functions too). We can improve the situation using inline functions. Inline annotation means that the specific function, along with the function parameters, will be expanded at the call site; this helps reduce call overhead.

Similarly, the inline keyword can be used with properties and property accessors that do not have the backing field. Let's see how in this recipe.

Getting ready

You need to install a preferred development environment that compiles and runs Kotlin. You can also use the command line for the purpose, for which you need Kotlin compiler installed along with JDK. I am using online IDE at `https://try.kotlinlang.org/` to compile and run my Kotlin code for this recipe. You can also use IntelliJ IDEA as the development environment.

How to do it...

Let's see how to work with inline properties in these steps:

1. Let's try an example where we `inline` an accessor of a property in Kotlin:

```
var x.valueIsMaxedOut: Boolean
inline get() = x.value == CONST_MAX
```

2. In this example, we just used the `inline` keyword with the `get` accessor. We can also declare both the `get` and `set` accessors as inline by making the whole property inline, as shown in this code snippet:

```
inline var x.valueIsMaxedOut: Boolean
get() = x.value == CONST_MAX
set(value) {
    // set field here
    println("Value set!")
}
```

In the preceding snippet, both accessors are inlined.

3. One thing to keep in mind, though, is that inline does not work with property or accessor if the property has a backing field or the assessor does not reference the backing field. The code here is an example of a scenario where we cannot use `inline`:

```
var x.valueIsMaxedOut: Boolean = true
get() = x.value == CONST_MAX
set(value) {
    // set field here
    println("Value set!")
}
```

Another thing to keep in mind is that, although inline properties reduce call overhead by getting expanded only at the call site, they also increase the overall bytecode, so inline should not be used with large functions or accessors.

How it works...

So, basically, we use inline when we wish to reduce memory overhead. Like the inline function, we can also declare properties as inline or the accessors of properties as inline. However, one thing to keep in mind is that inlining increases bytecode considerably, so it is suggested to not inline functions or accessors that have a large code logic.

How to work with nested class

In this recipe, we will see how to use nested classes in Kotlin. A nested class is a member of its enclosing class.

Getting ready

You need to install a preferred development environment that compiles and runs Kotlin. You can also use the command line for the purpose, for which you need Kotlin compiler installed along with JDK. I am using online IDE at https://try.kotlinlang.org/ to compile and run my Kotlin code for this recipe.

How to do it...

Now we will see how to work with a nested class in the following steps:

1. Let's try an example of a nested class in Kotlin:

```
fun main(args: Array<String>) {
    var a1 = outCl()
    a1.printAB()
    outCl.inCl().printB()
}
class outCl {
var a = 6
    fun printAB () {
    var b_ = inCl().b
```

```
            println ("a = $a and b = $b_ from inside outCl")
    }

    class inCl {
        var b = "9"
            fun printB() {
                println ("b = $b from inside inCl")
            }
        }
    }
```

Here's the output:

```
a = 6 and b = 9 from inside outCl
b = 9 from inside inCl
```

2. Now, let's try an example of the inner class. To declare a nested class as inner, we use the inner keyword. An inner class can access members of the outer class, as they carry a reference to the outer class:

```
fun main(args: Array<String>) {
    var a = outCl()
    a.printAB()
    a.inCl().printAB()
}
class outCl {
    var a = 6
    fun printAB () {
        var b_ = inCl().b
        println ("a = $a and b = $b_ from inside outCl")
    }
    inner class inCl {
        var b = "9"
        fun printAB() {
            println ("a = $a and b = $b from inside inCl")
        }
    }
}
```

The output of the preceding code is this:

```
a = 6 and b = 9 from inside outCl
a = 6 and b = 9 from inside inCl
```

How it works...

A nested class can be created by just declaring the nested class inside another class. In this case, to access the nested class, you make a static reference that is like `outerClass.innerClass()`, and you can also make an object of inner class using this. An `inner` class, on the other hand, is created by adding the `inner` keyword to a nested class. In that case, we access the inner class as though it was a member or the outer class, that is, using an object of the outer class like this:

```
var outerClassObject = outerClass()
outerClassObject.innerClass().memberVar
```

A nested class does not have access to members of the outer class, as it does not have any reference to an object of the outer class. On the other hand, the inner class can access all of the outer class's members, as it has a reference to an object of outer class.

There's more...

We can also create anonymous inner classes in Kotlin using the `object` keyword, like this:

```
val customTextTemplateListener = object:ValueEventListener{
    override fun onCancelled(p0: DatabaseError?) {
    }
    override fun onDataChange(dataSnapshot: DataSnapshot?) {
    }
}
```

Getting class in Kotlin

In this recipe, we will look into the ways by which we can get the class reference in Kotlin. Primarily, we will be working with reflection. Reflection is a library that provides the ability to inspect code at runtime instead of compile time. In Java, we can get a variable's class through `getClass()`, like `something.getClass()`. Let's see how to resolve a variable's class in Kotlin.

How to do it...

1. Java's equivalent of resolving a variable's name is with the `.getClass()` method, for example, `something.getClass()`. In Kotlin, we can achieve the same thing with `something.javaClass`.

2. To get a reference to the reflection class, we used to do `something.class` in Java, whose Kotlin equivalent is `something::class`. This returns a `KClass`. The special features of this `KClass` is that it provides introspection capabilities quite similar to the abilities provided to Java's reflection class.

   Note that the KClass is different from Java's `Class` object. If you want to obtain a Java `Class` object from Kotlin's `KClass`, use the `.java` extension property:

   ```
   val somethingKClass: KClass<Something> = Something::class
   val a: Class<Something> = somethingKClass.java
   val b: Class<Something> = Something::class.java
   ```

3. The latter example will be optimized by the compiler to not allocate an intermediate KClass instance.

   If you use Kotlin 1.0, you can convert the obtained Java class to a KClass instance by calling the `.kotlin` extension property, for example, `something.javaClass.kotlin`.

There's more...

As was just described, `KClass` provides you with introspection capabilities. Here are a few methods of `KClass`:

- `isAbstract`: True if this class is abstract
- `isCompanion`: True if this class is a companion object
- `isData`: True if this class is a data class
- `isFinal`: True if this class is final
- `isInner`: True if this class is an inner class
- `isOpen`: True if this class is open

Follow this link (`https://kotlinlang.org/api/latest/jvm/stdlib/kotlin.reflect/-k-class/`) for the complete list of functions provided by KClass.

Working with delegated properties

Kotlin 1.1 brought many updates; one of the important ones was delegated properties. There are three types of delegated properties:

- `lazy`: Lazy properties are the ones evaluated first and the same instance is returned after them, much like a cache
- `observable`: The listener is notified whenever a change is made
- `map`: Properties are stored in the map instead of in every field

In this recipe, we will see how to work with these delegates. So let's get started.

Getting ready

We will be working on Android code, so we will require Android Studio 3.

How to do it...

Let's see a simple example of a delegated property:

1. First, we will work with the lazy delegate property. Simply put, this delegate can suspend the object creation until we first access it. This is really important when you are working with heavy objects; they take a long time to be created—for example, when creating a database instance or maybe dagger components. Not only this, the result is remembered and the same value is returned for subsequent calls for `getValue()` on this kind of delegated property. Let's take a look at an example:

```
override fun onCreate(savedInstanceState: Bundle?) {
    super.onCreate(savedInstanceState)
    val button by lazy { findViewById<Button>(R.id.submit_button) }
}
setContentView(R.layout.activity_main)
    button.text="Submit"
}
```

2. The preceding is a standard `onCreate` method of an activity. If you look carefully, we have set the `button` variable before the `setContentView(..)` method. When you run it, it runs perfectly. If you hadn't used lazy, it would have given a `NullPointerException`, something like this:

```
Caused by: java.lang.NullPointerException: Attempt to invoke
virtual method 'void
android.widget.Button.setText(java.lang.CharSequence)' on a null
object reference
```

3. The button variable was null, as we called it before the `setContentView`. However, this wasn't a problem with lazily created `button` object, because although we had declared it before `setContentView`, the `button` object wasn't created. It was created after its first access, that is, when we tried to set a property on it.

4. So, with a lazy construct, you don't need to think about where to place your code for initialization, and initialization of an object is deferred till its first use.

Another key thing to note is that, by default, the evaluation of lazy properties will be synchronized, which means the value is computed in one thread, and the rest of the threads will see the same value. There are three types of initialization:

- `LazyThreadSafetyMode.SYNCHRONIZED`: This is the default mode and ensures that only a single thread can initialize the instance.
- `LazyThreadSafetyMode.PUBLICATION`: In this mode, multiple threads can execute the initialization.
- `LazyThreadSafetyMode.NONE`: This mode is used when we are sure that initialization will happen only on one thread. For example, in the case of Android, we can be sure that views will be initialized by the UI thread only. Since this doesn't guarantee thread-safety, it has much less overhead.

Another useful delegate is the observable delegate. This delegate helps us observe any changes to the property. For example, let's take a look at a very basic implementation of the `observable` delegate:

```
fun main(args: Array<String>) {
    val paris=Travel()
     paris.placeName="Paris"
     paris.placeName="Italy"
}
class Travel {
    var placeName:String by Delegates.observable("<>"){
```

```
            property, oldValue, newValue ->
                println("oldValue = $oldValue, newValue = $newValue")
        }
    }
```

This is the output:

```
oldValue = <>, newValue = Paris
oldValue = Paris, newValue = Italy
```

As we can see, the `observable` delegate takes in two things: a default value (which we specified as <>) and a handler, which gets called whenever that property is modified.

Let's now work with the `vetoable` delegate. It's a lot like the `observable` delegate, but with it, we can "veto" the modification. Let's look at an example:

```
fun main(args: Array<String>) {
    val paris=Travel()
    paris.placeName="Paris"
    paris.placeName="Italy"
    println(paris.placeName)
}
class Travel {
    var placeName:String by Delegates.vetoable("<>"){
        property, oldValue, newValue ->
            if(!newValue.equals("Paris")){
                return@vetoable false
            }
            true
    }
}
```

This is the output:

```
Paris
```

As you can see in the preceding example, if `newValue` isn't equal to `"Paris"`, we will return `false`, and the modification will be aborted. If you want the modification to take place, you need to return `true` from the construct.

Sometimes, you create an object based on values dynamically, for example, in the case of parsing JSON. For those applications, we can use the `map` instance itself as the delegate for a delegated property. Let's see an example here:

```
fun main(args: Array<String>) {
    val paris=Travel(mapOf(
        "placeName" to "Paris"
    ))
    println(paris.placeName)
}
class Travel(val map:Map<String,Any?>) {
    val placeName: String by map
}
```

Here's the output:

```
Paris
```

To make it work for `var` properties, you need to use a `MutableMap`, so the preceding example might look something like this:

```
fun main(args: Array<String>) {
    val paris=Travel(mutableMapOf(
        "placeName" to "Paris"
    ))
    println(paris.placeName)
}
class Travel(val map:MutableMap<String,Any?>) {
    var placeName: String by map
}
```

Of course, the output will be the same.

There's more...

The observable delegated property can be used extensively in adapters. Adapters are used to populate data in some sort of list. Usually, when data is updated, we just update the member variable list in the adapter and then call `notifyDatasetChanged()`. With the help of observable and `DiffUtils`, we can just update the things that are actually changed, rather than changing everything. This results in much more efficient performance.

Working with enums

Enums are used when a variable can only take one of a small set of possible values. An example would be the case of type constants (direction: "North", "South", "East", and "West"). With the help of enums, you can avoid errors from passing in invalid constants, and you also document which values are acceptable for use.

In this recipe, we will see how to use enums in Kotlin.

Getting ready

We'll be using an IntelliJ IDEA for writing and running the code. First, we will be creating a simple type-safe enum, `Direction`, with the members NORTH, SOUTH, EAST, and WEST (representing four directions).

How to do it...

Let's see an example of the `enum` class:

1. In this example, we will create an enum of directions. We will assume that there are only four of them:

```
enum class Direction {
    NORTH,SOUTH,EAST,WEST
}
fun main(args: Array<String>) {
    var north_direction=Direction.NORTH
    if(north_direction==Direction.NORTH){
        println("Going North")
    }else{
        println("No idea where you're going!")
    }
}
```

2. As you can see, the variable (`north_direction`) can just take values among the predefined constants in the `enum` class.

3. We can also initialize enums with default values:

```
enum class Direction(var value:Int) {
    NORTH(1),SOUTH(2),EAST(3),WEST(4)
}
fun main(args: Array<String>) {
    var north_direction=1
    if(north_direction==Direction.NORTH.value){
        println("Going North")
    }else{
        println("No idea where you're going!")
    }
}

//Output: Going North
```

There's more...

It is strongly recommended that you don't use enums in your Android projects. According to Google engineers, adding a single enum will increase the size to approximately 13 times that of the final DEX file. It also generates the problem of runtime overhead and your app will require more space.

The Android documentation says this:

> *"Enums often require more than twice as much memory as static constants. You should strictly avoid using enums on Android."*

However, if you want the comfort of enums, you can use Android's annotation library, which has `TypeDef` annotations—but sadly, this isn't supported by Kotlin at the time of this book being written, so we hope it gets added in future versions of Kotlin.

4
Functions

The following recipes will be covered in this chapter:

- Specifying default values in functions
- Using named arguments in functions
- Creating the `RecyclerView` Adapter in Kotlin
- Creating getter setters in Kotlin
- Passing variable arguments to a function
- Passing a function as a parameter to another
- Declaring a `static` function
- Using the `use` keyword in Kotlin
- Working with Closure in Kotlin
- Function literals with receivers
- Working with anonymous functions

Introduction

Functions are the building blocks of any code. They help make our programs more modular, secure, and easy to understand. Functions are crucial in object-oriented programming as they play an important role in abstraction and encapsulation (two very important design principles). Kotlin brings a lot of updates to the way we use functions. It makes functional programming easier with the help of chaining and lambdas. In this chapter, we will learn recipes that will help us work with functions. So let's get started!

Specifying default values in functions

If you come from the Java world, you might remember that we can't specify a default value to methods. This means that we can't do something like this in Java:

```
public void foo(int a, int b=10){
}
```

We need to write two methods for it, and it is known as *method overloading:*

```
public void foo(int a){
}

public void foo(int a, int b){
}
```

Also, suppose you have a function with three different kinds of parameters, such as these:

```
public void foo (int a,double b, String c){
}
```

Then you'll have seven instances of method overloading:

```
public void foo (int a,double b, String c),
public void foo (int a,double b) ,
public void foo (double b, String c),
public void foo (int a, String c),
public void foo (int a),
public void foo (double b),
public void foo (String c)
```

Kotlin provides you with default values in the methods by which you can prevent an insane amount of method overloading. Some people might say, "Hey, why don't we use the builder pattern instead of method overloading?". Those people are right, but Kotlin's way is easier than that. Let's see how!

Getting ready

We will be using IntelliJ IDEA to write and execute our code. You can use whatever development environment you are comfortable with.

How to do it...

In Kotlin, parameters of functions can have default values, and they are used when the corresponding argument is omitted. This, in turn, reduces the number of overloads. The preceding example with three different types of parameters can be resolved easily in Kotlin with a lot less code:

1. Let's add the mentioned code in the editor, run it, and check the output:

```
fun main(args: Array<String>) {
    foo()
    foo(1)
    foo(1,0.1)
    foo(1,0.1,"custom string")
}
fun foo(a:Int=0, b: Double =0.0, c:String="some default value"){
    println("a=$a , b=$b ,c = $c")
}
```

If you run the preceding code, you will see the following output:

```
Output:
a=0 , b=0.0 ,c = some default value
a=1 , b=0.0 ,c = some default value
a=1 , b=0.1 ,c = some default value
a=1 , b=0.1 ,c = custom string
```

2. As you can see, we didn't have to implement four different methods, and we could map the arguments. The default parameters are used when we don't call the methods by providing explicit parameters, so when you don't pass any parameters, it just uses all the default ones. With the help of named arguments, we can decrease the number of methods even further, but we will cover this in the next recipe.

3. One thing to note is that default arguments will also work with constructors. So you can have a class declaration as follows:

```
data class Event(var eventName: String? = "", var eventSchedule:
Date? = Date(), var isPrivate: Boolean = false)
```

 To learn more about a data class, head on to the Chapter 11, *How to create data class recipe.*

4. Then we can declare objects, as shown:

```
Event("Celebration")
Event("Ceberation",Date())
Event("Ceberation",Date(),true)
```

As you can see, with the help of default values in the constructors, we are avoiding the need to implement multiple constructors, which we used to do in Java.

Remember that there is a catch here. We won't be able to do this if you are creating objects in Java. This means that doing things as shown in the following code will not be accepted by Java. Now I know you'll be like "What happened to 100% interoperability with Java ?!":

```
new Event("Celebration")
new Event("Celebration",Date())
new Event("Celebration",Date(),true)
```

5. We just need to do a small modification if we want to expose multiple overloads to Java callers, that is—namely adding @JvmOverloads to the constructors and functions with default values so that the preceding class declaration becomes this:

```
data class Event @JvmOverloads constructor (var eventName: String?
= "", var date: Date? = Date(), var isPrivate: Boolean = false)
```

6. Also, our method becomes this:

```
@JvmOverloads fun foo(a:Int=0, b: Double =0.0, c:String="some
default value"){
 println("a=$a , b=$b ,c = $c")
 }
```

This is a small price to pay, but the @JvmOverloads annotation helps our constructors and functions to have default values, called from the Java world too.

There's more...

If we want our code to work only in the Kotlin world, then we don't need the @JvmOverloads annotation because Kotlin has its own rules by which it can work with default values in constructors and functions. Adding the @JvmOverloads annotation creates all the necessary overloads. So if you decompile your Kotlin bytecode, you will see all the overloaded versions of constructors and functions.

Using named arguments in functions

This recipe can be thought of as an extension to the previous recipe, *Specifying default values in functions*. Default parameters and named arguments in the function together can bring down the number of method overloads by a huge amount. We've already seen how to use default parameters in functions; now, let's see how to use name arguments.

Getting ready

We will be using IntelliJ IDEA to write and execute our code. You can use whatever development environment you are comfortable with.

How to do it...

Another step forward to reduce the number of overloads and increase code readability is to use named arguments. Let's take look at the following code:

1. Taking the same example of the `foo` function, here's how we can use named arguments:

```
fun main(args: Array<String>) {
    foo(b=0.9)
    foo(a=1,c="Custom string")
}
 fun foo(a:Int=0, b: Double =0.0, c:String="some default value"){
    println("a=$a , b=$b ,c = $c")
}
```

2. This is the output that you will get by running the preceding code:

```
Output:
a=0 , b=0.9 ,c = some default value
a=1 , b=0.0 ,c = Custom string
```

3. The named arguments prevent us from overloads and also make our code much more readable. Also, we don't need to put in all the arguments. What I mean is, if you just had two parameters—a and c—then you would have to do something like this:

```
foo(1, 0.0, "Custom string")
```

4. You have to add a default value to fill the space between a and c. However, with named arguments, you are able to use `foo(a=1,c="Custom string")` without needing default arguments in between.

5. One key thing to note is that when we call a function with both positional and named arguments, we need to place the positional arguments before the first named one. For example, the `foo(1,b = 0.1)` call is allowed, but `foo(a = 1, 0.1)` is not.

The default values and named arguments can bring down the number of overloads needed to a minimum, making the code size small and improving the code's readability.

Creating the RecyclerView Adapter in Kotlin

`RecyclerView` is among the most widely used elements in Android development. It is essentially used to display data in a list using an adapter. In this recipe, we will learn how to leverage great things in Kotlin to make `RecyclerView` much more efficient. We will also be using `DiffUtils`. It is available from 24.02. According to the documentation:

> *DiffUtil is a utility class that can calculate the difference between two lists and output a list of update operations that converts the first list into the second one.*

The definition is self-explainatory. The `notifyDatasetChanged` is a very expensive operation of the adapter. The `DiffUtils` only updates the parts that were changed, unlike `notifyDatasetChanged`, which updates the whole list.

Getting ready

Create a new Android project in Android Studio. You can also clone the `https://gitlab.com/aanandshekharroy/kotlin-cookbook` repository and check out the **1-recycler-view-in-kotlin** branch.

In this app, we will be creating a simple list of different Android flavors released by Google, something like what's seen here:

As you can see, there is a floating action button; clicking it will update the order of the list. We will be updating the list (`RecyclerView`), but we will update it using `DiffUtils` instead of the `notifyDatasetChanged` method.

How to do it...

So, let's now follow these steps to create the app we just discussed:

1. First, we need to create a list of Android flavors. So, we will first create a data class that takes in image and name of flavor:

```
data class AndroidFlavours (var name:String, val image:Int)
```

We have defined the type of image as `Int` because we will be using the IDs of drawable items. In the `drawable` folder, we will be keeping all the required images.

2. Next, we will create a list of Android flavors:

```
val flavorList= listOf<AndroidFlavours>(
        AndroidFlavours("Cupcake",R.drawable.cupcake),
        AndroidFlavours("Donut",R.drawable.donut),
        AndroidFlavours("Eclair",R.drawable.eclair),
        AndroidFlavours("Froyo",R.drawable.froyo),
        AndroidFlavours("Gingerbread",R.drawable.gingerbread),
        AndroidFlavours("HoneyComb",R.drawable.honeycomb),
        AndroidFlavours("Icecream Sandwich",R.drawable.icecream),
        AndroidFlavours("Jellybean",R.drawable.jellybean),
        AndroidFlavours("KitKat",R.drawable.kitkat),
        AndroidFlavours("Lollipop",R.drawable.lollipop))
```

3. Now, we will create an adapter. We will name it `AndroidFlavourAdapter`:

```
class
AndroidFlavourAdapter:RecyclerView.Adapter<AndroidFlavourAdapter.Fl
avourViewHolder>() {
    var flavourItems:List<AndroidFlavours> by
Delegates.observable(emptyList()){
        property, oldValue, newValue ->
        notifyChanges(oldValue,newValue)
    }

    override fun onCreateViewHolder(parent: ViewGroup, viewType:
Int): FlavourViewHolder {
        return
FlavourViewHolder(parent.inflate(R.layout.flavour_item))
    }

    override fun getItemCount(): Int =flavourItems.size

    override fun onBindViewHolder(holder: FlavourViewHolder,
position: Int) {
holder.name.text=flavourItems.get(holder.adapterPosition).name
holder.image.loadImage(flavourItems.get(holder.adapterPosition).ima
ge)
    }

    inner class FlavourViewHolder(var view:
View):RecyclerView.ViewHolder(view){
        var name:TextView = view.findViewById(R.id.textView)
        var image:ImageView = view.findViewById(R.id.imageView)
    }
}
```

The preceding code is quite standard for the general implementation of `RecyclerView`, except for the two things.

One of these is the `loadImage` function, which is not a native function but an extension function, whose implementation is this:

```
fun ImageView.loadImage(image: Int) {
    Glide.with(context).load(image).into(this)
}
```

4. Another thing is that we have defined the list of `AndroidFlavours` in the adapters. The `flavoursList` in the adapter is an `observable` property. This means the listener gets notified of changes to this property. Hence, we get the following construct:

```
var flavourItems:List<AndroidFlavours> by
Delegates.observable(emptyList()){
    property, oldValue, newValue ->
    notifyChanges(oldValue,newValue)
}
```

5. Now, whenever we try to assign a value to the `flavourItems` variable, the construct under the { .. } block is run, and we have old and new values to do an operation if we want. In this case, we will do it using the `notifyChanges` method. Let's look at the `notifyChanges` method:

```
private fun notifyChanges(oldValue: List<AndroidFlavours>,
newValue: List<AndroidFlavours>) {
    val diff = DiffUtil.calculateDiff(object : DiffUtil.Callback()
{
        override fun getChangePayload(oldItemPosition: Int,
newItemPosition: Int): Any? {
            val oldFlavor=oldValue.get(oldItemPosition)
            val newFlavor=newValue.get(newItemPosition)
            val bundle=Bundle()
            if(!oldFlavor.name.equals(newFlavor.name)){
                bundle.putString("name",newFlavor.name)
            }
            if(!oldFlavor.image.equals(newFlavor.image)){
                bundle.putInt("image",newFlavor.image)
            }
            if(bundle.size()==0) return null
            return bundle
        }

        override fun areItemsTheSame(oldItemPosition: Int,
```

```
newItemPosition: Int): Boolean {
        return
oldValue.get(oldItemPosition)==newValue.get(newItemPosition)
    }

    override fun areContentsTheSame(oldItemPosition: Int,
newItemPosition: Int): Boolean {
        return
oldValue.get(oldItemPosition).name.equals(newValue.get(newItemPosit
ion).name)&&oldValue.get(oldItemPosition).image.equals(newValue.get
(newItemPosition).image)
    }

    override fun getOldListSize() = oldValue.size

    override fun getNewListSize() = newValue.size

})

diff.dispatchUpdatesTo(this)
}
```

I will explain the preceding code in the next section.

6. Now, let's set up the adapter:

```
mAdapter= AndroidFlavourAdapter()
flavour_list.layoutManager=LinearLayoutManager(this)
flavour_list.adapter=mAdapter
mAdapter.flavourItems=flavorList
shuffle.setOnClickListener {
    mAdapter.flavourItems=flavorList.shuffle()
}
```

7. The `shuffle` function will just randomize the order of the list of `AndroidFlavours`. The `.shuffle()` function is not a native function provided by Kotlin or Java, but an extension function:

```
fun <E> List<E>.shuffle(): MutableList<E> {
    val list = this.toMutableList()
    Collections.shuffle(list)
    return list
}
```

How it works...

Let's dive into the `DiffUtils`. The `DiffUtils` requires two arrays/lists, one of which should be the old list and the other should be the new list.

There are five main functions:

- `getNewListSize()`: This returns the size of the new list.
- `getOldListSize()`: This method returns the size of the old list.
- `areItemsTheSame()`: This method is used to determine whether two objects represent the same item.
- `areContentsTheSame()`: This method is used to determine whether the two objects contain the same data. In our implementation, we are returning true if both objects have the same name and image.
- `getChangePayload()`: When `areItemsTheSame()` returns true and `areContentsTheSame()` returns false, then `DiffUtils` calls this method to get the payload of changes.

In our implementation of the preceding method, we are adding the change of name and image in the payload:

```
override fun getChangePayload(oldItemPosition: Int,
newItemPosition: Int): Any? {
    val oldFlavor=oldValue.get(oldItemPosition)
    val newFlavor=newValue.get(newItemPosition)
    val bundle=Bundle()
    if(!oldFlavor.name.equals(newFlavor.name)){
        bundle.putString("name",newFlavor.name)
    }
    if(!oldFlavor.image.equals(newFlavor.image)){
        bundle.putInt("image",newFlavor.image)
    }
    if(bundle.size()==0) return null
    return bundle
}
```

Finally, after the diff calculation, the `DiffUtils` object dispatches the changes to the Adapter. To do that, we call the `dispatchUpdatesTo` method:

```
diff.dispatchUpdatesTo(this)
```

To update the changes from the data in the payload, you need to override `onBindViewHolder (holder: FlavourViewHolder, position: Int, payloads: MutableList<Any>?)`:

```
override fun onBindViewHolder(holder: FlavourViewHolder, position:
Int, payloads: MutableList<Any>?) {
    if (payloads != null) {
        if (payloads.isEmpty())
            return onBindViewHolder(holder,position)
        else {
            val o = payloads.get(0) as Bundle
            for (key in o.keySet()) {
                if (key == "name") {
                    holder.name.text=o.getString("name")
                } else if (key == "image") {
                    holder.image.loadImage(o.getInt("image"))
                }
            }
        }
    }
}
```

The changes in the payload are dispatched using the `notifyItemRangeChanged` method of the adapter.

There's more...

The documentation states that the `DiffUtils` might take some time to process the diff between two lists if the lists are too big, so this must be calculated on a background thread, for example, using `RxJava`.

Creating getter setters in Kotlin

If you have worked with Java, you probably know what a *getter-setter* is. Java has fields and getter-setters are the methods that are used to **access** (getter) and **modify** (setter) member variables. They are an essential part of encapsulation (one of the design principles).

However, in Kotlin, we don't have any fields, but we have **properties** instead. A property can have a custom implementation of an **accessor** and a **mutator.** In this recipe, we will see how we can implement custom accessors and mutators.

Getting ready

We will be using IntelliJ IDEA to write and execute our code. You can use whatever development environment you are comfortable with. We will be using examples to understand the custom getter-setters of Kotlin.

How to do it...

Let's follow these steps to understand how custom getter-setters work in Kotlin:

1. The syntax of a Kotlin `property` looks like this:

```
var <propertyName>[: <PropertyType>] [= <property_initializer>]
[<getter>]   [<setter>]
```

So if you use something like `val a =1`, you get a default `getter` and `setter`.

2. Now, let's see how we can create a custom `getter`. Suppose we have a property whose value depends on another property:

```
fun main(args: Array<String>) {
    val sample=Sample()
    println(sample.isListBig)
}
class Sample{
    val array= mutableListOf<Int>(1,2,3)
    val isListBig:Boolean
        get()=array.size>2
}
```

If you run the preceding code, you'll see the output in the console as follows:

```
Run   HelloWorldKt
      /usr/lib/jvm/java-1.8.0-openjdk-amd64/bin/java ...
      true

      Process finished with exit code 0
```

3. As you can see, we can modify the getter in the `get` method of the property. If the property type is inferred from the getter, we can also do this:

```
val isListBig get()=array.size>2
```

The result will be the same, of course.

Now, let's take a look at accessors:

1. In Java, we used to do something like the following:

```
public setIsListBig(boolean isListBig){
    this.isListBig=isListBig
}
```

2. If we try to pull this off in Kotlin, it will look something like this:

```
        var isListBig :Boolean = false
            set(value) {
Recursive call    this.isListBig= array.size>2
                }
    }
```

3. As you can see, we will get a warning from IDE suggesting that it is a recursive call. Why? Because when you are trying to set a value using `.isListBig`, you are already using a setter inside a setter, hence the **recursive cycle**.

4. In order to get away from this recursive call and still implement a setter, you need the `field` keyword. So the preceding implementation will look something like this:

```
var isListBig :Boolean = false
    set(value) {
        field= array.size>2
    }
```

5. When you initialize `isListBig` while declaring the property, the value is assigned to the backing field without invoking the setter. The `field` keyword is used to access the backing field, and it will be generated for a property if it uses the default implementation of at least one of the accessors, or if a custom accessor references it through the `field` identifier.

6. If you want to restrict the access of your setter, you can do so with the following:

```
var isListBig :Boolean = false
    private set(value) {
        field= array.size>2
    }
```

7. Also, suppose you are using some sort of dependency injection. You can do it with this:

```
var mPresenter:MainActivityMvpPresenter?=null
    @Inject set
```

8. Similar to `set`, you can also have a custom implementation of `get`. Let's look at an example:

```
class SameClass {
    var name="aanand"
    get() = field.toUpperCase()
}
```

9. Now, let's say that we are trying to access the `name` property:

```
fun main(args: Array<String>) {
    var s=SameClass()
    println(s.name)
}
```

If you run the preceding code, you'll see this output:

Note that we have used `field` in the `get()` method too. It's the same backing field that we explained earlier.

There's more...

One thing to note here is that you cannot implement a custom getter or setter for your property in the constructor. You need to declare a property in the body of the class:

```
class Student(val name: String, age: Int) {
  var age: Int = age
      set(value) {
        println("Setting age to $value")
        field = value
    }
}
```

One key thing to note here is that you need to keep the visibility of the getter exactly similar to the visibility of the property:

```
protected var name="aanand"
protected get() = field.toUpperCase()
```

The preceding code is perfectly valid, though it's redundant to place the same access modifier again, hence it's better to omit it.

The setter, on the other hand, can have an access modifier less permissive than the property. Consider this example:

```
protected var name="aanand"
    private set
```

The preceding code is valid because the access modifier of the setter, `private`, is less permissive than the property's access modifier:

```
protected var name="aanand"
    public set
```

The preceding code, however, is not valid, as `protected` is less permissive than `public`.

Passing variable arguments to a function

There are a lot of scenarios in which we need to pass variable arguments to a function. In Kotlin, we can do that using the `vararg` modifier. In this recipe, we will go through all the ways of doing that. We will look at a few examples to demonstrate how to use this feature of Kotlin.

Getting ready

You need to install the preferred development environment that compiles and runs Kotlin. You can also use the command line for this purpose, for which you need Kotlin compiler installed, along with JDK. I am using an online IDE at https://try.kotlinlang.org/ to compile and run my Kotlin code for this recipe. You can also use IntelliJ IDEA for your development environment.

How to do it...

Let's go through the following steps, where we demonstrate how to pass a variable number of arguments to a function:

1. Using `vararg`, we can pass comma-separated arguments to a function, where we have defined the single argument to a method as `vararg`, as in the following example:

```
fun main(args: Array<String>) {
    someMethod("as","you","know","this","works")
}
fun someMethod(vararg a: String) {
    for (a_ in a) {
        println(a_)
    }
}
```

2. Also, if you already have an array of values, you can directly pass it using the * spread operator:

```
fun main(args: Array<String>) {
    val list = arrayOf("as","you","know","this","works")
    someMethod(*list)
}
fun someMethod(vararg a: String) {
    for (a_ in a) {
        println(a_)
    }
}
```

So basically, `vararg` tells the compiler to take the passed arguments and wrap them into an array.

3. The spread operator, on the other hand, simply tells the compiler to unwrap array members and pass them as separate arguments. The spread operator—that is, *–is put just before the name of the array being passed in.

4. However, obviously one may always need to pass other arguments, named arguments, and so on.
   In the following example code, we try to pass another argument other than vararg:

```
fun main(args: Array<String>) {
    val list = arrayOf("as","you","know","this","works")
    someMethod(3, *list)
}
fun someMethod(b: Int, vararg a: String) {
    for (a_ in a) {
        println(a_)
    }
}
```

5. In the next example, the first argument is similar to the vararg type, but it works:

```
fun main(args: Array<String>) {
    someMethod("3", "as","you","know","this","works")
}
fun someMethod(b: String, vararg a: String) {
    println("b: " + b)
    for (a_ in a) {
        println(a_)
    }
}
```

The output is as follows:

```
b: 3
as
you
know
this
works
```

6. So usually, vararg is the last argument passed, but what if we want to pass other arguments after vararg? We can, but they have to be named. That is why the following code will not compile:

```
// does not compile
```

```
fun main(args: Array<String>) {
    someMethod("3", "as","you","know","this","works", "what")
}
fun someMethod(b: String, vararg a: String, c: String) {
    println("b: " + b)
    for (a_ in a) {
        println(a_)
    }
    println("c: " + c)
}
```

7. It does not compile because the last string passed in it is considered part of vararg, and the compiler throws an error because we did not pass the value of c. To do it correctly, we need to pass c as a named argument, just as shown here:

```
fun main(args: Array<String>) {
    someMethod("3", "as","you","know","this","works", c = "what")
}
fun someMethod(b: String, vararg a: String, c: String) {
    println("b: " + b)
    for (a_ in a) {
        println(a_)
    }
    println("c: " + c)
}
```

The output is as follows:

```
b: 3
as
you
know
this
works
c: what
```

How it works...

The vararg modifier tells the compiler to take all comma-separated arguments and wrap them into an array, while *—that is the spread operator—unwraps elements of the array and passes them as arguments.

There's more...

What if we want the first argument to have a default value, like in this example:

```
fun main(args: Array<String>) {
    someMethod("3", "as","you","know","this","works")
}
fun someMethod(b: String = "x", vararg a: String) {
    println("b: " + b)
    for (a_ in a) {
        println(a_)
    }
}
```

We want all arguments to be considered as part of `vararg`, but the compiler reads the first argument as `b`. In this case, naming the passed arguments can solve the problem:

```
fun main(args: Array<String>) {
    someMethod(a = *arrayOf("3", "as","you","know","this","works"))
}
fun someMethod(b: String = "x", vararg a: String) {
    println("b: " + b)
    for (a_ in a) {
        println(a_)
    }
}
```

In the preceding code, the compiler understands that the value of `b` is not passed, and it takes the default value. Similarly, if you want to have two `vararg` in your function, you will need to pass named arguments.

Passing a function as a parameter to another

Kotlin gives us the power to declare *high-order functions*. In a high-order function, we can pass and return functions as parameters. This is an extremely useful feature and makes our code much more easy to work with. In fact, many of the Kotlin library's functions are high order, such as `map`. In Kotlin, we can declare functions and function references as values that are then passed in to the function. In this section, we will first understand how to declare lambdas and then how to pass them into a function.

Getting ready

You need to install a preferred development environment that compiles and runs Kotlin. You can also use the command line for this purpose, for which you need Kotlin compiler installed, along with JDK. I am using an online IDE at `https://try.kotlinlang.org/` to compile and run my Kotlin code for this recipe. You can also use IntelliJ IDEA as the development environment.

How to do it...

Let's follow these steps to understand the working of high-order functions:

1. Let's start by understanding how we declare functions as lambdas:

```
fun main(args: Array<String>) {
    val funcMultiply = {a:Int, b:Int -> a*b}
    println(funcMultiply(4,3))
    val funcSayHi = {name: String -> println("Hi $name")}
    funcSayHi("John")
}
```

2. In the preceding code block, we declared two lambdas: one (`funcMultiply`) that takes two integers and returns an integer, and another (`funcSayHi`) lambda that takes a string and returns a unit—that is, it returns nothing.

3. Although we did not need to declare the type of arguments and the return type in the preceding example, in some cases we need to explicitly declare the argument types and return types. We do this in the following way:

```
fun main(args: Array<String>) {
    val funcMultiply : (Int, Int)->Int = {a:Int, b:Int -> a*b}
    println(funcMultiply(4,3))
    val funcSayHi : (String)->Unit = {name: String -> println("Hi $name")}
    funcSayHi("John")
}
```

4. So now that we have a general idea of how lambdas work, let's try and pass one in another function—that is, we will try a high-order function. Check out this code snippet:

```
fun main(args: Array<String>) {
    val funcMultiply : (Int, Int)->Int = {a:Int, b:Int -> a*b}
    val funcSum : (Int, Int)->Int = {a:Int, b:Int -> a+b}
```

```
        performMath(3,4,funcMultiply)
        performMath(3,4,funcSum)
}
fun performMath(a:Int, b:Int, mathFunc : (Int, Int) -> Int) : Unit
{
        println("Value of calculation: ${mathFunc(a,b)}")
}
```

5. Yup, it is as simple as that—create a function lambda and pass it into the function. So this is just one aspect of a high-order function—that is, we can pass a function as an argument to the function.

6. Another use of high-order functions is to return a function. Consider the following example where we need a function that transforms the total price of an order according to certain conditions. Kind of like in an e-commerce site, but way simpler:

```
fun main(args: Array<String>) {
    val productPrice1 = 600; // free delivery of order above 499
    val productPrice2 = 300; // not eligible for free deliver
    val totalCost1 = totalCost(productPrice1)
    val totalCost2 = totalCost(productPrice2)

    println("Total cost for item 1 is
${totalCost1(productPrice1)}")
    println("Total cost for item 2 is
${totalCost2(productPrice2)}")
}
fun totalCost(productCost:Int) : (Int) -> Int{
    if(productCost > 499){
        return { x -> x }
    }
    else {
        return { x -> x + 50 }
    }
}
```

7. Note how we need to change functions that we apply based on certain conditions so that we return a function that suits the conditions. We assign the returned function to a variable and then we can just put append () in front of the variable to use it as a function, just like we did with the lambdas. This works because the high-order function is essentially returning a lambda.

How it works...

In Kotlin, we can assign a function to a variable, and then we can pass it into a function or return it from a function. This is because it's essentially declared like a variable. This is done using a lambda declaration of functions.

Declaring a static function

Static functions are very useful as they help us prevent copying the same methods in multiple objects so you can follow the **don't repeat yourself** (**DRY**) rule. They are also useful when you don't need to create an instance of an object. In Kotlin, we don't have static methods/functions and variables, like we did in Java, but we can still achieve the same results. Let's see how!

Getting ready

We will be using IntelliJ IDEA to write and execute our code. You can use whatever development environment you are comfortable with. We will be learning about static functions by going through the examples and their workings.

How to do it...

One of the use cases of static methods is that we can prevent multiple copying of the same methods in different classes, and also that we don't need to create an object of the enclosing class.

Kotlin recommends creating package-level functions. If you are coming from the Java world, this probably won't make any sense to you as this isn't supported in Java. Let's see how it's done in Kotlin:

1. You need to create a Kotlin file with the `.kt` extension and just declare the method that you'll be using in many places. I have created a `SampleClass.kt` file and have added a method that we will be calling from other classes:

```
package packageA
fun foo(){
    println("calling from boo method")
}
```

2. Now, I'll call this method from `HelloWorld.kt`:

```
import packageA.*
fun main(args: Array<String>) {
    foo()
}
```

3. Since the function was present in `packageA`, we used the `import` statement. This way, we followed DRY and didn't need to create an instance of any class.

4. Another way to do it is by putting methods or variables in an object declaration. So we can modify the `SameClass.kt` into the following:

```
package packageA
object Foo{
    fun callFoo() = println("Foo")
    var foo="foo"
}
```

5. Any method or variable under object declaration will work as a `static` method or variable. In order to access it, we can do this:

```
Foo.callFoo()
```

This is much like how we call static methods.

6. However, suppose you want the class name as a qualifier and access elements of the class. You can still use it using the `companion` keyword. Here's how it will look:

```
fun main(args: Array<String>) {
    SampleClass.foo()
}
class SampleClass{
    companion object {
        fun foo()= print("In foo method")
    }
}
```

7. If you want to call the method under the `companion` object, you'll need to access it like this:

```
SampleClass.Companion.foo();
```

8. If `Companion` seems like an eyesore to you, you can use the `@JvmStatic` annotation:

```
companion object {
    @JvmStatic
    fun foo()= print("In foo method")
}
```

9. Then, you can access it using `SampleClass.foo()`, just like you do in a Kotlin class.

Using the use keyword in Kotlin

There are some situations where if you use a resource (for example, a file) then you have to take care of its lifecycle so that you don't leak resources. For example, if you read from a file, you need to close it after use, or else you'll leave it in an unstable state. Java 7 brought an update that could handle this without a need to handle it explicitly. Kotlin also provides this feature, but in a much easier way. It does so by using the `use` method. We will learn about this in the following recipe. So let's get started!

Getting ready

We will be using IntelliJ IDEA to write and execute our code. You can use whatever development environment you are comfortable with.

How to do it...

Let's take the following steps to understand the `use` function of Kotlin:

1. To understand the `use` keyword, we will need to go back to Java. Prior to Java 7, managing the resources that needed to be closed was a bit cumbersome. For example, look at the following code:

```
private static void printFile() throws IOException {
    InputStream input = null;

    try {
        input = new FileInputStream("sampleFile.txt");
        // Some operation using input object
```

```
    } finally {
        if(input != null){
            input.close();
        // closing the resource
        }
    }
}
```

2. Let's examine the preceding code. We know an exception can be thrown inside the `try` block when we use the `input` object. However, it can also be thrown in the `finally` block, because we are trying to close the `input` object. Now, the `finally` block will be called whether or not the `try` block throws an exception. Suppose both the `try` and `finally` blocks throw exceptions—which one of the two will propagate? The answer is that the exception will be thrown in the `finally` block, even if the exception of `try` would make more sense here.

3. Java 7 brought an update to this problem by introducing the try-with-resource construct, which looks something like this:

```
try(FileInputStream input = new FileInputStream("file.txt")) {
    int data = input.read();
    // operations on input object
}
```

4. When the `try` block finishes executing, the `FileInputStream` object is closed automatically. Also, if both the operations—`input.read()` and the closing of the input object—throw exceptions, the exception thrown by `input.read()` will propagate. The `use` keyword of Kotlin does the exact same work. In this section, we will see how.

5. In the preceding example, we saw in Java will look something like the following in Kotlin if we implement the `use` keyword:

```
FileInputStream("file.txt").use {
    input ->
    var data = input.read()
}
```

How it works...

`use` accepts a function literal and is defined as an extension on an instance of closeable. It will close down the resource, just like the try-with-resources construct, after the function has completed, whether an exception was raised or not.

Working with closures

MDN (`https://developer.mozilla.org/en-US/docs/Web/JavaScript/Closures`) says this:

> *"A closure is a special kind of object that combines two things: a function, and the environment in which that function was created. The environment consists of any local variables that were in-scope at the time the closure was created"*

Closures in functional programming are the functions that are *aware* of their surroundings. By this, I mean that a closure function has access to the variables and parameters defined in the outer scope. Remember that in Java and traditional procedural programming, the variables were tied to the scope, and as soon as the block got executed, local properties were blown out of the memory. Java 8 lambdas can access outer variables, but can't modify them, and this limits the capabilities if you try to do functional programming in Java 8. Let's take a look at an example where we work with closures in Kotlin.

Getting ready

We will be using IntelliJ IDEA for writing and executing our code. You can use whatever development environment you are comfortable with.

How to do it...

In this example, we will simply create an array of integers and calculate its sum:

```
fun main(args: Array<String>) {
    var sum=0
    var listOfInteger= arrayOf(0,1,2,3,4,5,6,7)
    listOfInteger.forEach {
        sum+=it
    }
    println(sum)
}
```

In the preceding example, the `sum` variable is defined in the outer scope; still, we are able to access and modify it.

There's more...

If you want to learn more about high-order functions or closures, head on to the *Passing a function as a parameter to another* recipe of this chapter.

Function literals with receivers

A **Function literal** is a function that is not declared but that is passed in as an expression. Lambdas and anonymous functions are function literals. In Kotlin, we can call a function literal with a receiver object, and we can call methods on the receiver object inside the body of the function literal, quite like extension functions. In this recipe, we will learn how to use function literals with receivers.

Getting ready

You need to install a preferred development environment that compiles and runs Kotlin. You can also use the command line for this purpose, for which you need Kotlin compiler installed, along with JDK. I am using an online IDE at `https://try.kotlinlang.org/` to compile and run my Kotlin code for this recipe. You can also use IntelliJ IDEA for development environment.

How to do it...

Follow these steps to understand function literals:

1. Let's start with a simple function literal on a `String`, which returns a string added to the receiver string:

```
fun main(args: Array<String>) {
    var str1 = "The start of a "
    val addStr = fun String.(successor: String): String {
        return this + successor
    }
    str1 = str1.addStr("beautiful day.")
    println(str1)
}
```

A function literal has access to the receiver it has been called on, and it can access methods associated with that receiver.

2. We can also pass the receiver as a parameter in an ordinary function, where the first parameter is for a receiver. This can be useful in scenarios where we need to use an ordinary function.
   So `String.(String) -> Int` is similar to `(String, String) -> Int` is compatible. Check out the following example :

```
fun main(args: Array<String>) {
    var str1 = "The start of a "
    val addStr = fun String.(successor: String): Int {
        return this.length + successor.length
    }
    var x = str1.addStr("beautiful day.")
    println(x)
    fun testIfEqual(op: (String, String) -> Int, a: String, b:
String, c: Int) =
    assert(op(a, b) == c)

    testIfEqual(addStr, "The start of a ", "beautiful day.",
str1.length + "beautiful    day.".length) // OK
}
```

If the receiver type can be inferred, then lambda can be used as the function literal.

So basically, we can call a function literal on a receiver object, and inside the body of the function, we can access and call methods on a receiver object, similar to an extension function in Kotlin. The following is the syntax for this:

```
receiver.functionLliteral(arguments) -> ReturnType
```

Working with anonymous functions

In Kotlin, we can have functions as expressions by creating lambdas. Lambdas are function literals—that is, they are not declared as they are expressions and can be passed as parameters. However, we cannot declare return types in lambdas. Although the return type is inferred automatically by Kotlin compiler in most cases, for cases where it cannot be inferred on its own or it needs to be declared explicitly, we use anonymous functions. In this recipe, we will see how to use anonymous functions.

Getting ready

You need to install a preferred development environment that compiles and runs Kotlin. You can also use the command line for this purpose, for which you need Kotlin compiler installed, along with JDK. I am using an online IDE at https://try.kotlinlang.org/ to compile and run my Kotlin code for this recipe. You can also use IntelliJ IDEA for the development environment.

How to do it...

In the following steps, we will learn about anonymous functions with the help of some examples:

1. Let's start by declaring a function as a lambda:

```
fun main(args: Array<String>) {
    val funcMultiply = {a:Int, b:Int -> a*b}
    println(funcMultiply(4,3))
    val funcSayHi = {name: String -> println("Hi $name")}
    funcSayHi("John")
}
```

In the preceding code block, we have declared two lambdas: one (funcMultiply) that takes two integers and returns an integer, and another (funcSayHi) lambda that takes a string and returns a unit—that is, it returns nothing.

2. Although in the preceding example we did not need to declare the type of arguments and return type, in some cases we need to explicitly declare the argument types and return types. We do that in the following way, by means of an anonymous function:

```
fun main(args: Array<String>) {
    var funcMultiply = fun (a: Int, b: Int): Int {return a*b}
    println(funcMultiply(4,3))
    fun(name: String): Unit = println("Hi $name")
}
```

3. So now we have a general idea of how anonymous functions work. Now, let's try and pass one in another function—that is, we will try a high-order function. Check out this code snippet:

```
fun main(args: Array<String>) {
    var funcMultiply = fun(a: Int, b: Int): Int { return a*b }
    var funcSum = fun(a: Int, b: Int): Int { return a+b }
    performMath(3,4,funcMultiply)
    performMath(3,4,funcSum)
}
fun performMath(a:Int, b:Int, mathFunc : (Int, Int) -> Int) : Unit
{
    println("Value of calculation: ${mathFunc(a,b)}")
}
```

4. So basically, an anonymous function is declared just like a regular function, but without a name. The body can be an expression, as in the following example, or a block, as in the preceding example. One thing to note is that parameters are always passed inside the parentheses in the case of anonymous functions, unlike in lambda expressions:

```
fun main(args: Array<String>) {
    performMath(3,4,fun(a: Int, b: Int): Int = a*b )
    performMath(3,4,fun(a: Int, b: Int): Int = a+b )
}
fun performMath(a:Int, b:Int, mathFunc : (Int, Int) -> Int) : Unit
{
    println("Value of calculation: ${mathFunc(a,b)}")
}
```

5. Another interesting difference between a lambda and an anonymous function is that in a lambda, the return statement returns from the enclosing function, whereas in an anonymous function, it simply returns from the function itself.

6. One can omit the parameter type and return type from an anonymous function as well if it can be inferred on its own.

7. Anonymous functions can access and modify variables inside their closures.

So basically, one can declare an anonymous function just like a regular function without a name (hence, the name anonymous). It can be an expression or a code block.

5

Object-Oriented Programming

The following recipes will be covered in this chapter:

- Working with interfaces in Kotlin
- How to implement complicated interfaces with multiple overridden methods in Kotlin
- How to extend a class in Kotlin (Inheritance and Extension functions)
- How to work with Generics in Kotlin
- How to implement polymorphism in Kotlin
- Restricting class hierarchies

Introduction

Object-Oriented Programming, also known as **OOP**, is a programming paradigm based on objects. In this programming paradigm, objects contain data in the form of fields and code in the form of methods, which can be used to modify the data of the same object. In some object-oriented languages, objects are instances of classes (for example, Java and Kotlin). In object-oriented programming, our code is made up of objects that interact with each other. In this chapter, we will learn about some key components of OOPs, such as interfaces, classes, class hierarchies, and Generics.

Working with interfaces in Kotlin

Interfaces in OOP are like the *contract*. They define the behavior or rules. The classes that implement them need to do so in order to conform to the behavior defined by interfaces. However, that's not it. Interfaces in Kotlin provide much more. Prior to Java 8, we couldn't have the implementation of methods in the interfaces, but in Kotlin, we can have that too! In this recipe, we will see how to deal with interfaces in Kotlin.

Getting ready

I'll be using IntelliJ IDEA for writing and executing code. You are free to use any IDE where you can run the Kotlin code.

How to do it...

As we have just discussed, interfaces in Kotlin can have the implementation of methods; let's follow these mentioned steps to check that out:

1. Let's create an interface named `DemoInterface`:

    ```
    interface DemoInterface {

        fun implementatedMethod() {
            println("From demo interface")
        }
    }
    ```

    Defining a method with implementation in the interface is just like you would do inside a class.

2. Now, let's see a class that has implemented the preceding interface:

    ```
    class IntefaceImplementation: DemoInterface
    ```

3. Then, you can call the method like this:

    ```
    fun main(args: Array<String>) {
        var interfaceImplementation= IntefaceImplementation()
        interfaceImplementation.implementatedMethod()
    }
    ```

Here's the output:

```
From demo interface
```

4. A key benefit to this new type of interface is that you can have the behavior of multiple interfaces since it allows method implementation:

```
fun main(args: Array<String>) {
    var interfaceImplementation= IntefaceImplementation()
    interfaceImplementation.foo()
    interfaceImplementation.bar()

}
interface A {

    fun foo() {
        println("foo from A")
    }
}
class IntefaceImplementation: A,B

interface B  {
    fun bar() {
        println("foo from B")
    }
}
```

As you can see in the preceding code, using multiple interfaces, we have the behavior of two entities. Yes, this may sound like multiple inheritance.

5. Suppose you have two types of interfaces, and both have the methods with the same name, as follows:

```
interface A {
    fun foo() {
        println("foo from A")
    }
}
interface B  {
    fun foo() {
        println("foo from B")
    }
}
```

6. Now, if you try to implement both interfaces to a class, the compiler will throw an error:

```
Error:(24, 1) Kotlin: Class 'IntefaceImplementation' must override
public open fun foo(): Unit defined in packageB.A because it
inherits multiple interface methods of it
```

7. The reason is intuitive, as it brings ambiguity of which method to call. Hence, Kotlin will require you to implement that method and call the desired method inside it, something like this:

```
class IntefaceImplementation: A,B {
    override fun foo() {
        super<A>.foo()
        super<B>.foo()
    }
}
```

8. Now, you will simply call the `foo` method, as earlier:

```
fun main(args: Array<String>) {
    var interfaceImplementation= IntefaceImplementation()
    interfaceImplementation.foo()
}
```

This is the output:

```
foo from A
foo from B
```

Interfaces in Kotlin can have the implementation of methods, but can't have states. This means you can't declare a property in the interface and store the state in it. Either the class implementing it needs to override it, or you need to implement its accessor also.

For example, you can't have `val a=23` in an interface, though you can have something like the following:

```
val a: Int
    get() = 2
```

Alternatively, simply define it in the interface and override it in the implementing class, like this:

```
class InterfaceImplementation: A,B {
    override val a: Int=25}
```

Next, we will look at interfaces delegation in Kotlin:

1. A delegation pattern, an object (`https://en.wikipedia.org/wiki/Object_` `(computer_science)`) handles a request by delegating to a second object. Let's take a look at the following code:

```
fun main(args: Array<String>) {
    var interfaceImplementation= InterfaceImplementation(object :A{
    })
    interfaceImplementation.someMethod()
}
class InterfaceImplementation(var a:A){
    fun someMethod(){
        a.foo()
    }
}
interface A {
    fun foo() {
        println("foo from A")
    }
}
```

2. In the preceding example, we are delegating the call to the `foo` method, to the object that has implemented the interface A. While the preceding code is perfectly fine, Kotlin allows us to use the function directly. Look at this code:

```
class InterfaceImplementation(var a:A):A by a{
    fun someMethod(){
        foo()
    }
}
```

3. As you can see, the `InterfaceImplementation` class is implementing A but is delegating the implementation to the objects that it is receiving as parameters.

There's more...

Now that Kotlin supports the implementation of methods in interfaces, you might be thinking what's the difference between an `interface` and `abstract` methods.

In interfaces, you can only define the property, which needs to be overridden by implementing class. However, in an abstract class, you can have an implementation that works with the state so that it cannot be overridden in the derived classes. In an abstract class, you can define some states and methods that will be the same in the derived class.

Another key difference is that you can have final members in an abstract class, but not in interfaces. Also, interfaces don't support `protected` and `internal` modifiers. It only supports `private`.

How to implement complicated interfaces with multiple overridden methods in Kotlin

SOLID is a mnemonic acronym that is used to define the five basic **object-oriented design** principles:

- Single Responsibility Principle
- Open-Closed Principle
- Liskov Substitution Principle
- Interface Segregation Principle
- Dependency Inversion Principle

The **Interface Segregation Principle(ISP)** states that if an interface becomes too long, it is better to split it into smaller pieces (interfaces) so that the client doesn't need to implement the ones in which they are not interested. In this recipe, we will understand what and why this is important.

Getting ready

We will be using Android Studio 3.0. Ensure that you have its latest version.

How to do it...

Let's see an example where ISP can help us:

1. This is a simple example of a "fat" interface:

```
button.setOnClickListener(object : View.OnClickListener {
    fun onClick(View v) {
        // TODO: do some stuff...
    }
    fun onLongClick(View v) {
        // we don't need it
    }

    fun onTouch(View v, MotionEvent event) {
        // we don't need it
    }
});
```

2. As you can see, the problem of a big interface is that we are forced to implement the methods even if we don't have anything to do it in there.

3. A simple solution is to break that interface into smaller interfaces, like the following code:

```
interface OnClickListener {
    fun onClick( v:View )
}
public interface OnLongClickListener {
    fun onLongClick( v: View)
}
interface OnTouchListener {
    fun onTouch( v: View,  event: MotionEvent)
```

4. Note that now we have divided the one big interface into smaller ones, which can be implemented independently.

5. Kotlin also has a powerful feature that allows you to write full implementation of methods in the interfaces itself. Let's take a look at the following code to understand it:

```
fun main(args: Array<String>) {
    Simple().callMethod()
}
class Simple:A{
    fun callMethod(){
        bar()
    }
}
interface A{
    fun bar(){
        println("Printing from interface")
    }
}
```

6. As you can see, we implemented the whole method in the interface, and we were able to call it from the class that implemented that interface.

7. This feature can also be used to follow the ISP principle, as we can put a commonly used method in the interfaces itself; as a result, we will not need to implement it every time we implement that interface.

How to extend a class in Kotlin (Inheritance and Extension functions)

In this recipe, we will learn how to extend a class (Inheritance) and how to extend the functionality of a class using Kotlin's Extension functions.

Inheritance is probably the first concept you learn in object-oriented programming. It is a mechanism where a new class is derived from an existing class. Via this, the classes may inherit or acquire the properties and methods of other classes. **Extension functions**, on the other hand, let us skip creating wrapper for functionality and enable us to add extra functions to the classes. Let's see both of them now.

Getting ready

Since we will be dealing with Android code, it is recommended that you use Android Studio as IDE. The source code can be found in the **1-recycler-view-in-kotlin** branch of the `https://gitlab.com/aanandshekharroy/kotlin-cookbook` repository.

How to do it...

A class derived from another class is called a subclass, whereas the class from which a subclass is derived is called a superclass. In this example, we will create a superclass A and a subclass B. To extend class B, we need to use : in the class declaration, and then add the superclass name with its primary constructor. Let's take a look at the following steps:

1. A key thing to remember is that classes in Kotlin are *closed* for extension by default, so we need to open them by adding the *open* keyword before the class declaration. So our superclass A looks like this:

   ```
   open class A
   ```

2. Then, we can extend our class B, as follows:

   ```
   class B:A()
   ```

3. Now, suppose our class A has a primary constructor that takes in a `String` variable, such as this:

   ```
   open class A(var str:String)
   ```

   Now, if we wish to extend B with A, there are two ways to do so:

   - Initialize A in B's primary constructor. In this approach, we will initialize A by passing arguments from B's primary constructor. Consider this example:

     ```
     class B(var randomString:String) : A(randomString)
     ```

- If B, or any class, doesn't have a primary constructor, then each secondary class of the extending class needs to initialize the superclass using the `super` keyword. Consider the given example:

```
class B: A{
    constructor(randomString:String) : super(randomString)
    constructor(randomString:String, randomInt:Int) :
super(randomString)
```

4. We generally extend a class to import the functionalities from a superclass and sometimes, we might also like to override them to have our own implementation. Similar to classes, methods are also closed by default, and we need to "open" them with the open modifier:

```
open class A(var str:String){
    open fun foo(){
        println("foo from A")
    }
}
class B(var string: String): A(string) {
    override fun foo(){
        println("foo from B")
    }
}
```

5. You can also mark a method "final" to prevent any other subclass from overriding it. Take this example into consideration:

```
open class A(var str:String){
    final fun foo(){
        println("foo from A")
    }
}
```

6. If you extend your class with an abstract class, you need to implement all the methods defined as abstract in the abstract class. Note that you don't need to mark them open in order for them to be overridden by the extending class. Making them abstract does the job itself, as shown in this example:

```
class B(var string: String): C() {
    override fun methodC() {
        // Do something here
    }
}
abstract class C{
    abstract fun methodC()
```

```
        fun impl(){}
    }
```

Extension functions

Extension functions are useful, as they allow us to extend the functionality of a class without actually touching it. For example, if you've used Glide or Picasso library for placing an image inside the `Imageview`, you must be familiar with the following code:

```
Glide.with(context).load(image_url).into(imageView)
```

We can make this look much better using an extension function. Let's call the `loadImage(imageUrl)` function on `imageView`. If you do it, you will see an error—**Unresolved reference- loadImage**:

You will also see two suggestions, one of which is **Create extension function**:

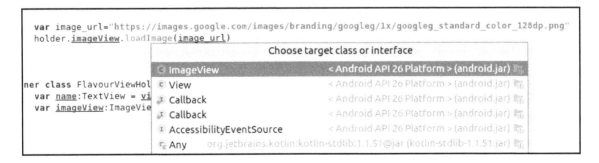

If you click on **Create extension function**, you'll be provided with some choices, as in the this screenshot:

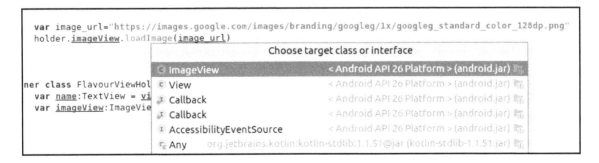

Click on **ImageView**, because we want to create an extension function on it.

When you click on it, an extension function is created in the same file, which looks like this:

```
private fun ImageView.loadImage(image_url: String) {
}
```

Here, we can place our Glide/Picasso code for image loading:

```
private fun ImageView.loadImage(image_url: String) {
    Glide.with(context).load(image_url).into(this)
}
```

So, even if the `loadImage` function was not present in the `ImageView` class, we were able to extend it and use it as if this function was part of `ImageView`, and we didn't even have to touch the `ImageView` class. The extension function extended the functionality of `ImageView` externally.

How it works...

The prefix to the extension function (the name that comes before the dot) is called **receiver type**, that is, the type being extended. This receiver object is accessed inside the function using the `this` keyword. The extension functions are resolved statically; it's like calling a static method. Since this is a static method, it's not needed to be defined under the class, but since it's a static method, it's hard to test. For example, Mockito (a testing framework) cannot test static methods, so to produce great code quality, use extension function only if that function doesn't require any testing.

There's more...

What happens when you create an extension function with a name similar to that of a member function? For example, in the following code, what will happen if we call `c.foo()`?:

```
fun main(args: Array<String>) {
    var c= C()
    c.foo()
}
class C{
    fun foo(){
        println("from member")
    }
```

```
}
private fun C.foo() {
    println("from extension")
}
```

This is the output we get:

from member

So a member function will win if an extension function with the same name is called.

How to work with Generics in Kotlin

Generic methods and classes help us use the same method or class for various types. This improves code reusability. In this recipe, we will understand Generics and how to use it in Kotlin. Generics in Kotlin is quite similar to Generics in Java, but there are additional special keywords in Kotlin that make Generics in Kotlin more intuitive. Let's dive in.

Getting ready

You need to install a preferred development environment that compiles and runs Kotlin. You can also use the command line for the purpose, for which you need a Kotlin compiler installed along with JDK. I am using IntelliJ IDE to compile and run my Kotlin code for this recipe.

How to do it...

Now, let's follow these steps to understand how Generics works in Kotlin, with the help of some examples:

1. Let's start with a generic class that we can instantiate using any type of parameter:

```
fun main(args: Array<String>) {
    val intgen: GenCl<Int> = GenCl<Int>(10)
    println(intgen.a)

    // We are letting Kotlin compiler infer type
    val strgen = GenCl("A string")
    println(strgen.a)
}
```

```
class GenCl<T>(t: T) {
    var a = t
}
```

The output of this program is this:

```
10
A string
```

2. We can also restrict what types are allowed in our generic class like this:

```
fun main(args: Array<String>) {
    val intgen: GenCl<Int> = GenCl<Int>(10)
    println(intgen.a)

    val flgen = GenCl(1.0)
    println(flgen.a)
}

// Restricting T to only be of type Number
class GenCl<T: Number>(t: T) {
    var a = t
}
```

3. If we try to use the preceding class with a type that is not a `Number`, for example, a `String`, we get the following error:

```
Error:(8, 17) Type parameter bound for T in constructor GenCl<T :
Number>(t: T)
 is not satisfied: inferred type String is not a subtype of Number
```

4. Now, let's try an example of a generic method:

```
fun main(args: Array<String>) {
    fun <T> addTwo(a: List<T>) {
        for(x in a) {
            println(x)
        }
    }

    addTwo(listOf(10,20,30,40))
    addTwo(listOf("a","b","c","d","e"))
}
```

The output of the preceding code will be as follows:

```
10
20
30
40
a
b
c
d
e
```

There's more...

Generic types in Java are invariant, which means `List<String>` is not a subtype of `List<Object>`. Java has this so that we are not able to add, say, a `Float` to a `List` that contains `String` and has the type as `Object`. In Kotlin, we have a better solution where we use the wildcard argument as `? extends E`, which denotes that the method accepts a subtype of E or collection of E and not just E itself. This gives us the power to read from a collection of E but not write to it, as we do not know what items are accepted. This makes Kotlin covariant.

How to implement polymorphism in Kotlin

Polymorphism is the ability of an object to take many forms, depending on the situation. Kotlin supports two types of polymorphism: **compile-time polymorphism** and **run-time polymorphism**. In this recipe, we will try both. Let's get started.

Getting ready

You need to install a preferred development environment that compiles and runs Kotlin. You can also use the command line for the purpose, for which you need a Kotlin compiler installed along with JDK. I am using IntelliJ IDE to compile and run my Kotlin code for this recipe.

How to do it...

In the following steps, we will learn how to use polymorphism in Kotlin:

1. Let's start with compile-time polymorphism. In **compile-time polymorphism**, the name functions, that is, the signature remains the same but parameters or return type is different. At compile time, the compiler then resolves which functions we are trying to call based on the type of parameters and more. Check out this example:

```
fun main(args: Array<String>) {
    println(doubleOf(4))
    println(doubleOf(4.3))
    println(doubleOf(4.323))
}

fun doubleOf(a: Int): Int {
    return 2*a
}

fun doubleOf(a: Float): Float {
    return 2*a
}

fun doubleOf(a: Double): Double {
    return 2.00*a
}
```

Here's the output of the preceding code:

```
8
8.6
8.646
```

2. Now, let's talk about run-time polymorphism. In **run-time polymorphism**, the compiler resolves a call to overridden/overloaded methods at runtime. We can achieve run-time polymorphism using method overriding. Let's try an example where we extend a superclass and override one of its member methods:

```
fun main(args: Array<String>) {
    var a = Sup()
    a.method1()
    a.method2()

    var b = Sum()
    b.method1()
```

```
        b.method2()
    }

open class Sup {
    open fun method1() {
        println("Printing method 1 from inside Sup")
    }

    fun method2() {
        println("Printing method 2 from inside Sup")
    }
}

class Sum: Sup() {
    override fun method1() {
        println("Printing method 1 from inside Sum")
    }
}
```

The output of the preceding code is this:

```
Printing method 1 from inside Sup
Printing method 2 from inside Sup
Printing method 1 from inside Sum
Printing method 2 from inside Sup
```

Here, the compiler resolves, at run-time, which method to execute.

Restricting class hierarchies

In this recipe, we will learn how to restrict the class hierarchies in Kotlin. Before we move ahead, let's understand why this is a cause worth spending our time on.

Getting ready

I'll be using Android Studio to run the code described in this recipe.

How to do it...

When we are sure that a value or a class can have only a limited set of types or number of subclasses, that's when we try to restrict class hierarchy. Yes, this might sound like an enum class but, actually, it's much more than that. Enum constant exists only as a single instance, whereas a subclass of a sealed class can have multiple instances that can contain state. Let's look at an example in the mentioned steps:

1. We will create a **sealed** class named `ToastOperation`. Under the same source file, we will define a `ShowMessageToast` subclass:

   ```
   class ShowMessageToast(val message:String):ToastOperation()
   ```

2. Also, we'll define a `ShowErrorToast` object:

   ```
   object ShowErrorToast:ToastOperation()
   ```

3. As you may have noted, I have defined an **object** rather than a full class declaration, because the `ShowErrorToast` **object** doesn't have any state. Also, by doing so, we have removed *is* from the *when* block, since there is just one instance.

   Now, we can use it in a `when` statement, as follows:

   ```
   fun doToastOperation(toastOperation: ToastOperation){
       when(toastOperation){
           is ShowMessageToast
   ->Toast.makeText(this,toastOperation.message,Toast.LENGTH_LONG).sho
   w()
           ShowErrorToast->Toast.makeText(this,"Error..
   Grr!",Toast.LENGTH_LONG).show()
       }
   }
   ```

4. The key benefit is that we don't need to implement the `else` block, which acted as the default block when the other statements didn't fit the bill.

According to documentation, a `sealed` class can have subclasses, but all of them must be declared in the same file as the sealed class itself. However, the subclasses of subclasses need not be defined in the same file. It is abstract by itself, and you cannot instantiate objects from it.

Here's our structure of `sealed` classes:

```
sealed class ToastOperation {
}
object ShowErrorToast:ToastOperation()
class ShowMessageToast(val message:String):ToastOperation()
```

As you can see, we've kept all the subclasses under the same source file in which we have defined the **sealed** class.

How it works...

In the preceding example, we were sure that we can only have two types of toasts: an error toast and a toast with a custom message. So we created a **sealed** class `ToastOperation` and created two subclasses of `ToastOperation`. Note that if we aren't sure of types of subclasses, we will not use a **sealed** class, in that case, an **enum** class might be better suited.

There's more...

If you are using Kotlin versions prior to 1.1, you'll need to implement the subclasses inside the sealed class, much like this:

```
sealed class ToastOperation {
    object ShowErrorToast:ToastOperation()
    class ShowMessageToast(val message:String):ToastOperation()
}
```

Note that you can use the preceding way in the new version of Kotlin as well.

6
Collections Framework

The following recipes will be covered in this chapter:

- How to merge two collections

- Splitting original collection into pair of collections

- Sorting a list by specified comparator

- Sorting in descending order

- Parsing a JSON response using Gson

- How to filter and map using lambda expressions

- How to sort a list of objects and keep null objects at the end

- How to implement a lazy list in Kotlin

- How to pad a string in Kotlin

- How to flatten an array or map

- How to sort collection by multiple fields in Kotlin

- How to use limit in Kotlin list

- How to create a 2D array in Kotlin

- How to skip the first N entries in Kotlin

Introduction

Collections framework is useful when we want to process items in a collection. If you have worked with Java, you are probably familiar with collections framework. The most common use of collections framework are maps, sets, lists, and so on. Kotlin too has its collection framework, but it's much better than Java's collection framework because, in Kotlin, we can leverage the functional programming approach to make our code more concise and easy to work with. So, let's dive into the recipes related to Kotlin's collection framework.

How to merge two collections

In this recipe, we will see how to merge two or more collections into one. However, before we move ahead, we need to understand the difference between mutable and immutable types. An immutable type object is an object that cannot be changed. For example, if we define an immutable list, we won't be able to add other objects to it. With that in mind, let's start the recipe!

Getting ready

I'll be using IntelliJ IDEA for coding. You can use whichever IDE you like as long as it is able to compile and run Kotlin code.

How to do it...

You can create a list in Kotlin with the `listOf` method. However, the list returned by this method is an immutable list, so we need to create a mutable list in order to add objects to it. Let's check out the mentioned steps:

1. Let's create two lists, `listA` and `listB`, as follows:

```
var listA= mutableListOf<String>("a","a","b")
var listB= mutableListOf<String>("a","c")
```

 If the type declaration is inferred from the objects inside the `listOf/mutableListOf` method, we won't need to declare the type declaration explicitly. So, the preceding code will be rewritten as `mutableListOf("a","a","b")`.

2. Now, we will try to add the contents of listA in listB. For that purpose, we will require the addAll() method:

```
fun main(args: Array<String>) {
    val listA= mutableListOf<String>("a","a","b")
    val listB= mutableListOf<String>("a","c")
    listB.addAll(listA)
    println(listB)
}
```

Here's the output:

```
[a, c, a, a, b]
```

3. Another way to merge two lists is using union. This returns the unique elements of the combined collection:

```
fun main(args: Array<String>) {
    val listA= mutableListOf<String>("a","a","b")
    val listB= mutableListOf<String>("a","c")
    val listC=listB.union(listA)
    println(listC)
}
```

This is the output:

```
[a, c, b]
```

4. Similarly, mutable sets can be merged too, the only difference is that addAll in a set will be similar to what we will receive with the union method; since it's a set, only a unique value is allowed:

```
val setA= mutableSetOf<String>("a","b","c")
val setB= mutableSetOf<String>("a","b","c","d")
setB.addAll(setA)
println(setB)
println(setB.union(setA))
```

This is the output:

```
[a, b, c, d]
[a, b, c, d]
```

If you want to merge two maps, you will need the `putAll()` method, as `addAll` and `union` are not present for the `map`:

```
val mapA= mutableMapOf<String,Int>("a" to 1, "b" to 2)
val mapB= mutableMapOf<String,Int>("a" to 2, "d" to 4)
mapA.putAll(mapB)
println(mapA)
```

Here's the output:

```
{a=2, b=2, d=4}
```

Notice that key `a` was defined in both the maps, but the one that comes later (in this case, `mapB`) wins.

Splitting original collection into pair of collections

There are times when you wish that you could just split a list into sublists without going into the `for` and `while` loops. Kotlin provides you with a function just for this occasion. In this recipe, we will see how to split a list based on some criteria.

Getting ready

I'll be using IntelliJ IDEA for writing and running Kotlin code; you are free to use any IDE that can do the same task.

How to do it...

Kotlin provides a `partition` function. According to the documentation of the partition function it does the following:

> *Splits the original array into a pair of lists, where the first list contains elements for which predicate yielded true, while the second list contains elements for which predicate yielded false.*

Let's understand it more clearly by going through this example:

1. In this example, we will create a list of numbers, and we want to split this list into two sublists: one having odd numbers and the other having even numbers:

```
fun main(args: Array<String>) {
    val listA= listOf(1,2,3,4,5,6)
    val pair=listA.partition {
        it%2==0
    }
    println(pair)
}
```

This is the output:

```
([2, 4, 6], [1, 3, 5])
```

2. As you can see in the preceding example, we need to put the condition in the predicate inside the partition block. The returned object is a Pair object, holding the two sublists.

3. The partition function also works with the set collection in a similar way:

```
val setA= setOf(1,2,3,4,5,6)
val pair=setA.partition {
    it%2==0
}
println(pair)
```

Here's the output:

```
([2, 4, 6], [1, 3, 5])
```

How it works...

Let's take a look at the implementation of the partition function in Kotlin:

```
public inline fun <T> Iterable<T>.partition(predicate: (T) -> Boolean):
Pair<List<T>, List<T>> {
    val first = ArrayList<T>()
    val second = ArrayList<T>()
    for (element in this) {
        if (predicate(element)) {
            first.add(element)
        } else {
            second.add(element)
```

```
        }
    }
    return Pair(first, second)
}
```

As you can see, the `partition` function is just an abstraction, that saves you from writing long for loops, but internally it does it the same old way.

There's more...

The `partition` function works in a similar way with arrays as well. Here are its different usages. Each of them works similarly, just producing lists of different types:

```
// Produces two lists
inline fun <T> Array<out T>.partition(
    predicate: (T) -> Boolean
): Pair<List<T>, List<T>>

// Breaks original list of Byte and produces two lists of Byte
inline fun ByteArray.partition(
    predicate: (Byte) -> Boolean
): Pair<List<Byte>, List<Byte>>


// Breaks original list of Short and produces two lists of Short
inline fun ShortArray.partition(
    predicate: (Short) -> Boolean
): Pair<List<Short>, List<Short>>

// Breaks original list of Int and produces two lists of Int
inline fun IntArray.partition(
    predicate: (Int) -> Boolean
): Pair<List<Int>, List<Int>>

// Breaks original list of Long and produces two lists of Long
inline fun LongArray.partition(
    predicate: (Long) -> Boolean
): Pair<List<Long>, List<Long>>

// Breaks original list of Float and produces two lists of Float
inline fun FloatArray.partition(
    predicate: (Float) -> Boolean
): Pair<List<Float>, List<Float>>

// Breaks original list of Double and produces two lists of Double
inline fun DoubleArray.partition(
```

```
    predicate: (Double) -> Boolean
): Pair<List<Double>, List<Double>>

// Breaks original list of Boolean and produces two lists of Boolean
inline fun BooleanArray.partition(
    predicate: (Boolean) -> Boolean
): Pair<List<Boolean>, List<Boolean>>

// Breaks original list of Char and produces two lists of Char
inline fun CharArray.partition(
    predicate: (Char) -> Boolean
): Pair<List<Char>, List<Char>>
```

Sorting a list by specified comparator

Sorting a list is one of the most common operations done on the list. When we try to sort a list of custom objects, we need to specify the comparator. Let's see how we can sort a list by the specified comparator.

Getting ready

I'll be using IntelliJ IDEA for writing and running Kotlin code; you are free to use any IDE that can do the same task.

How to do it...

In the following examples, we will try to sort objects based on certain properties. This will give us an idea of how to sort based on the specified comparator:

1. Let's create a `Person` class with age property. We will be sorting a list of person objects based on age:

```
fun main(args: Array<String>) {
    val p1=Person(91)
    val p2=Person(10)
    val p3=Person(78)
    val listOfPerson= listOf(p1,p2,p3)
    var sortedListOfPerson=listOfPerson.sortedBy {
        it.age
    }
}
```

```
class Person(var age:Int)
```

2. To sort a list based on the specified comparator, we need to use the `sortedBy` function:

```
fun main(args: Array<String>) {
    val p1=Person(91)
    val p2=Person(10)
    val p3=Person(78)
    val listOfPerson= listOf(p1,p2,p3)
    var sortedListOfPerson=listOfPerson.sortedBy {
        it.age
    }
}
class Person(var age:Int)
```

3. Kotlin also provides a `sortedWith` method, where you can specify your own implementation of comparator:

```
fun main(args: Array<String>)
{
  val p1=Person(91)
  val p2=Person(10)
  val p3=Person(78)
  val listOfPerson= listOf(p1,p2,p3)
  var sortedListOfPerson=listOfPerson
  .sortedWith<Person>(object:Comparator<Person>{
      override fun compare(p0: Person, p1: Person):Int {
        if(p0.age>p1.age){
            return 1
        }
        if(p0.age==p1.age){
            return 0
        }
        return -1
      }
  })
}
class Person(var age:Int)
```

How it works...

The `sortedBy` function is syntactic sugar provided by Kotlin. Internally, it's calling the `sortedWith` method that takes in a comparator.

Now, let's see the implementation of the `sortBy` function:

```
public inline fun <T, R : Comparable<R>> Iterable<T>.sortedBy(crossinline
selector: (T) -> R?): List<T> {
    return sortedWith(compareBy(selector))
}
```

The `sortBy` function calls the `sortedWith` method inside it, which is as following:

```
public fun <T> Iterable<T>.sortedWith(comparator: Comparator<in T>):
List<T> {
    if (this is Collection) {
        if (size <= 1) return this.toList()
        @Suppress("UNCHECKED_CAST")
        return (toTypedArray<Any?>() as Array<T>).apply {
sortWith(comparator) }.asList()
    }
    return toMutableList().apply { sortWith(comparator) }
}
```

Sorting in descending order

In the last recipe, we saw how to sort a list with a specified comparator. We provide the comparator, and it sorts it accordingly. Interestingly, Kotlin also provides a method to sort items of a list in descending order. In this recipe, we will see how to sort a collection of primitive objects as well as custom objects in descending order. So let's get started!

Getting ready

I'll be using IntelliJ IDEA for writing and running Kotlin code; you are free to use any IDE that can do the same task.

How to do it...

We will now see how to sort in descending order using some examples:

1. First, we will try to sort a simple list of integers:

```
val listOfInt= listOf(1,2,3,4,5)
var sortedList=listOfInt.sortedDescending()
sortedList.forEach {
```

```
    print ("${it} ")
}
```

This is the output:

5 4 3 2 1

2. Now, let's use our list of `Person` from the preceding recipe. To sort it in descending order, this is what we will do:

```
val p1=Person(91)
val p2=Person(10)
val p3=Person(78)
val listOfPerson= listOf<Person>(p1,p2,p3)
val sortedListOfPerson=listOfPerson.sortedByDescending {
    it.age
}
sortedListOfPerson.forEach {
    print ("${it.age} ")
}
```

Here's the output:

91 78 10

How it works...

The `sortedByDescending` works a bit like `sortedBy`. Internally, both use the `sortedWith` function:

```
public inline fun <T, R : Comparable<R>>
Iterable<T>.sortedByDescending(crossinline selector: (T) -> R?): List<T> {
    return sortedWith(compareByDescending(selector))
}
```

The following is the implementation of `compareByDescending`:

```
@kotlin.internal.InlineOnly
public inline fun <T> compareByDescending(crossinline selector: (T) ->
Comparable<*>?): Comparator<T> =
        Comparator { a, b -> compareValuesBy(b, a, selector) }
```

Note that just the order of variables is reversed to produce the descending order.

Parsing a JSON response using Gson

In this recipe, we will learn how to parse JSON. JSON is the most widely used data type for API responses. We will be using Gson, an open source library by Google. It's fast, and it scales very well even with a huge response.

Getting ready

I'll be using Android Studio for this purpose, and JSONObject is provided by Android SDK. We will be using Gson for JSON parsing. You can add it to your project by adding the following lines to your `build.gradle` file:

```
compile 'com.google.code.gson:gson:2.8.0'
```

How to do it...

Now, let's follow these steps to parse JSON data using Gson. For example, we will use a raw string here to keep things simple:

1. First, we will create dummy JSON data using a raw string, as follows:

```
val jsonStr="""
    {
     "name": "Aanand Shekhar",
     "age": 21,
     "isAwesome": true
    }
""".trimIndent()
```

2. Next, we will create a data class to hold this data. Here's how our data class looks:

```
data class Information(val name:String,val age:Int, val
isAwesome:Boolean)
```

3. Finally, we will use `Gson` to parse the JSON string:

```
val information:Information=
Gson().fromJson<Information>(jsonStr,Information::class.java)
```

Now you can use it just like a Kotlin object.

There's more...

You can create the data class automatically using some Android Studio plugins. One of the most widely used plugins is **RoboPOJOGenerator** (`https://github.com/robohorse/` `RoboPOJOGenerator`).

How to filter and map using lambda expressions

In this recipe, we will learn how to transform a list using a `map` function in Kotlin, and how to filter the list with whichever criteria we like. We will be using lambda functions, which provide a great way to do functional programming. So let's get started.

Getting ready

I'll be using IntelliJ IDEA for writing and running Kotlin code; you are free to use any IDE that can do the same task.

How to do it...

First, let's see how to use the **filter** function on a list. The filter function returns a list containing all elements matching the given predicate. We will create a list of numbers and filter the list based on even or odd.

The `filter` method is good for an immutable collection as it doesn't modify the original collection but returns a new one. In the filter method, we need to implement the predicate. The predicate, like the condition, is based on the list that is filtered.

For example, we know that even items will follow `it%2==0`. So the corresponding filter method will look like this:

```
val listOfNumbers=listOf(1,2,3,4,5,6,7,8,9)
var evenList=listOfNumbers.filter {
    it%2==0
}
println(evenList)

//Output: [2, 4, 6, 8]
```

Another variant of the filter function is `filterNot`, which, as the name suggests, returns a list containing all elements not matching the given predicate.

Another cool lambda function is `map`. It transforms the list and returns a new one:

```
val listOfNumbers=listOf(1,2,3,4,5,6,7,8,9)
var transformedList=listOfNumbers.map {
    it*2
}
println(transformedList)

//Output: [2, 4, 6, 8, 10, 12, 14, 16, 18]
```

A variant of the `map` function is `mapIndexed`. It provides the index along with the item in its construct:

```
val listOfNumbers=listOf(1,2,3,4,5)
val map=listOfNumbers.mapIndexed { index, it
    -> it*index}
println(map)

//Output: [0, 2, 6, 12, 20]
```

How to sort a list of objects and keep null objects at the end

We have already seen how to sort a list based on a specified parameter using a comparator. However, so far, we have worked with lists having non-null values. In this recipe, we will see how to sort a list of objects, which have the null property (on which we are sorting). So let's get started.

Getting ready

I'll be using IntelliJ IDEA for writing and running Kotlin code; you are free to use any IDE that can do the same task.

How to do it...

Now, let's follow these steps to sort a list, while keeping null objects at the end:

1. Let's create a `Person` class having an age property that can be null:

   ```
   class Person(var age:Int?)
   ```

2. Now, let's create a list of `Person` objects:

   ```
   val listOfPersons=listOf(Person(10), Person(20), Person(2),
   Person(null))
   ```

3. Finally, we want to sort them in ascending order, while keeping the null items at the end:

   ```
   val
   sortedList=listOfPersons.sortedWith(compareBy(nullsLast<Int>(),{it.
   age}))
   sortedList.forEach {
       print(" ${it.age} ")
   }
   ```

   The output is as follows:

   ```
   2 10 20 null
   ```

How it works...

We have used the `sortedWith` method. According to documentation, `sortedWith` does this:

> *Returns a sequence that yields elements of this sequence sorted according to the specified comparator.*

Apart from that, we've made use of the `kotlin.comparisons` package, which provides us two main functions used in the preceding solution:

- `public inline fun <T: Comparable<T>> nullsLast()`: This method provides a comparator of nullable comparable values considering null value greater than any other value. That's how we can get the null items at the end, because they are considered bigger than any other values.

- `compareBy(comparator: Comparator<in K>, crossinline selector: (T) -> K)`: This function accepts a comparator (such as `nullsLast()`) and a function that provides values for the comparator, and then combines them into a new comparator.

How to implement a lazy list in Kotlin

If the value of an element or expression is not evaluated when it's defined, but rather when it is first accessed, it is said to be **lazily evaluated**. There are many situations where it comes in handy. For example, you might have a list A and you want to create a filtered list from it, let's call it list B. If you do something like the following, the filter operation will be performed during the declaration of B:

```
val A= listOf(1,2,3,4)
var B=A.filter {
    it%2==0
}
```

This forces the program to initialize B as soon as it is defined. While this may not be a big deal for a small list, it can cause latency with bigger objects. Also, we can delay the object creation until we first need it. In this recipe, we will learn how we can implement a lazy list.

Getting ready

I'll be using IntelliJ IDEA for writing and running Kotlin code; you are free to use any IDE that can do the same task.

How to do it...

To create a lazy list, we need to convert the list into a sequence. A sequence represents lazily evaluated collections. Let's understand it with an example:

1. In the given example, let's first filter a list based on elements being odd or even:

```
fun main(args: Array<String>) {
    val A= listOf(1,2,3,4)
    var B=A.filter {
        println("checking ${it}")
        it%2==0
    }
```

```
}
```

This is the output:

```
checking 1
checking 2
checking 3
checking 4
```

In the preceding example, the filter function was evaluated only when the object was defined.

2. Now, let's convert the list into a sequence. Converting the list to a sequence is just one step away; you can convert any list to a sequence using the .asSequence() method, or by Sequence{ createIterator() }:

```
fun main(args: Array<String>) {
    val A= listOf(1,2,3,4).asSequence()
    var B=A.filter {
        println("checking ${it}")
        it%2==0
    }
}
```

3. If you run the preceding code, you won't see any output in the console, because the object hasn't been created yet. It will be created when list B is first accessed:

```
fun main(args: Array<String>) {
    val A= listOf(1,2,3,4).asSequence()
    var B=A.filter {
        println("checking ${it}")
        it%2==0
    }
    B.forEach {
        println("printing ${it}")
    }
}
```

```
//Output:checking 1
         checking 2
         printing 2
         checking 3
         checking 4
         printing 4
```

The `filter` function was evaluated when the items were accessed. This is called **lazy evaluation**.

How it works...

Sequences in Kotlin are potentially unbounded and are used when the length of the list is not known in advance (much like Streams in Java 8). Since it can be infinite, lazy evaluation is needed for this type of structure. Consider this example:

```
val seq= generateSequence(1){it*2}
seq.take(10).forEach {
    print(" ${it} ")
}
```

Here, `generateSequence` generates a sequence of infinite numbers, but when we call `take(10)`, only 10 items are evaluated and printed.

How to pad a string in Kotlin

Sometimes, to keep up with the length of the string, we pad the string with some characters. In many communication protocols, keeping the standard length of the payload is vital. Kotlin makes it very easy to pad the string with any character and length. Let's see how to use it.

Getting ready

I'll be using IntelliJ IDEA for writing and running Kotlin code; you are free to use any IDE that can do the same task.

How to do it...

In this recipe, we will use the `kotlin.stdlib` library of Kotlin. Specifically, we will be working with the `padStart` and `padEnd` functions. Let's now follow the given steps to understand how to use these functions:

1. Let's see an example of the `padStart` function:

```
fun main(args: Array<String>) {
    val string="abcdef"
    val pad=string.padStart(10,'-')
    println(pad)
}
```

This is the output:

```
----abcdef
```

2. Next, we look at an example of `padEnd`:

```
val string="abcdef"
val pad=string.padEnd(10,'-')
println(pad)
```

Here's the output:

```
abcdef----
```

How it works...

The padding functions are needed to expand the string to a certain length using the character provided with the function. So, if the length of the padded string is less than the string itself, it will just return the same string.

Another key thing to note is that by default, the padding character is a space character. This is the implementation of the `padStart` function:

```
public fun String.padStart(length: Int, padChar: Char = ' '): String
        = (this as CharSequence).padStart(length, padChar).toString()
```

As you can see, the default value for `padChar` is a space character, and it is called on a String object.

How to flatten an array or map

In the previous few recipes of this chapter, we learned how to create multidimension arrays. In this recipe, we will see how we can convert them to a 1D list, or *flatten* it.

Getting ready

I'll be using IntelliJ IDEA for writing and running Kotlin code; you are free to use any IDE that can do the same task.

How to do it...

We will be using the `.flatten` method of the `kotlin.stdlib` library. It takes in an array or collection and returns a single list of all elements from all collections/arrays in the given collection/array.

For example, with an array of arrays:

`[[1,2,3],[1,2,3],[1,2,3]] -> [1,2,3,1,2,3,1,2,3]`

```
fun main(args: Array<String>) {
    val a= arrayOf(arrayOf(1,2,3),arrayOf(1,2,3),arrayOf(1,2,3))
    a.flatten().forEach { print(" ${it} ") }
}

//Output:  1 2 3 1 2 3 1 2 3
```

For example, with a list of lists:

`[[1,2,3],[1,2,3],[1,2,3]] -> [1,2,3,1,2,3,1,2,3]`

```
fun main(args: Array<String>) {
    val a= listOf(listOf(1,2,3),listOf(1,2,3),listOf(1,2,3))
    a.flatten().forEach { print(" ${it} ") }
}
```

How it works...

Let's take a look at the implementation of the `flatten()` function:

```
public fun <T> Iterable<Iterable<T>>.flatten(): List<T> {
    val result = ArrayList<T>()
    for (element in this) {
        result.addAll(element)
    }
    return result
}
```

As you can see, it's just adding the items from iterables (array or lists) in a new list and returning that list.

How to sort collection by multiple fields in Kotlin

In this recipe, we will learn how to sort a collection by multiple fields in Kotlin. This often comes in handy when we want to give precedence to an object over another object when both have equal value on a specific property. For example, we might have a list of `Student` objects and want to arrange them in ascending order of age, but if two students have the same age, we will order them based on their GPA. In this recipe, we will see how to handle use cases like this. So let's get started!

Getting ready

I'll be using IntelliJ IDEA for writing and running Kotlin code; you are free to use any IDE that can do the same task.

How to do it...

Now, let's follow these steps to sort based on multiple fields of an object:

1. First, let's create the `Student` class:

```
class Student(val age:Int, val GPA: Double)
```

2. Then, create a list of `Student` objects:

```
val studentA=Student(11,2.0)
val studentB=Student(11,2.1)
val studentC=Student(11,1.3)
val studentD=Student(12,1.3)
val
studentsList=listOf<Student>(studentA,studentB,studentC,studentD)
```

3. To sort it on multiple fields, we simply need to do this:

```
val
sortedList=studentsList.sortedWith(compareBy({it.age},{it.GPA}))
```

4. If we print it now, we will get the following output:

```
sortedList.forEach {
    println("age: ${it.age}, GPA: ${it.GPA} ")
```

```
    }

    //Output: age: 11, GPA: 1.3
             age: 11, GPA: 2.0
             age: 11, GPA: 2.1
             age: 12, GPA: 1.3
```

How it works...

We have used the `sortedWith` function, which takes in a comparator. The comparator is provided by the `compareBy` function. The `compareBy` has an overload that takes multiple functions:

```
public fun <T> compareBy(vararg selectors: (T) ->
Comparable<*>?):Comparator<T>
```

As you can see in the preceding code, `vararg` allows us to take multiple functions in its construct and returns a comparator, which feeds the `sortedWith` function.

Note that sorting with multiple fields works like sort by field 1, then by field 2, then by field 3, and so on.

How to use limit in Kotlin list

In this recipe, we will learn how to take specific items from the list. We will use the `kotlin.stdlib` library for this purpose.

Getting ready

I'll be using IntelliJ IDEA for writing and running Kotlin code; you are free to use any IDE that can do the same task.

How to do it...

We will be using the `take` function and its variants for limiting the items in the list.

`take(n)`: Returns a list of the first n items:

```
fun main(args: Array<String>) {
    val list= listOf(1,2,3,4,5)
    val limitedList=list.take(3)
    println(limitedList)
}

//Output: [1,2,3]
```

`takeLast(n)`: Returns a list containing the last [n] elements:

```
fun main(args: Array<String>) {
    val list= listOf(1,2,3,4,5)
    val limitedList=list.takeLast(3)
    println(limitedList)
}

//Output: [3,4,5]
```

`takeWhile{ predicate }`: Returns a list containing the first elements satisfying the given [predicate]:

```
val list= listOf(1,2,3,4,5)
val limitedList=list.takeWhile { it<3 }
println(limitedList)

//Output: [1,2]
```

`takeLastWhile{predicate}`: Works a bit like `takeWhile`, except that it evaluates the list from the end.

`takeIf { predicate }`: Returns `this` value if it satisfies the given [predicate], or `null` if it doesn't:

```
fun main(args: Array<String>) {
    val list= listOf(1,2,3,4,5)
    var limitedList=list.takeIf { it .contains(1) }
    println(limitedList)
}

//Output: [1,2,3,4,5]
```

Note that the `it` inside the `takeIf` lambda represents the list itself, and not just an element of list.

How to create a 2D array in Kotlin

2D arrays are useful for data representation in certain situations such as board games, images, and so on. In Java, we can represent a 2D array by doing the following:

```
int[][] data = new int[size][size];
```

Since Kotlin brings new syntax, let's see how to work with a 2D array in Kotlin.

Getting ready

I'll be using IntelliJ IDEA for writing and running Kotlin code; you are free to use any IDE that can do the same task.

How to do it...

Let's now follow the given steps to create a 2D array in Kotlin:

1. We can create a simple 2D array in Kotlin using this syntax:

   ```
   val array = Array(n, {IntArray(n)})
   ```

   Here, n represents the dimension of the array. Here, we have used the Array class of Kotlin, which represents an array (specifically, a Java array when targeting the JVM platform). We have initialized the Array object by passing the size and initializer:

   ```
   public inline constructor(size: Int, init: (Int) -> T)
   ```

2. Our dimension is n, and as an initializer, we are passing a 1D array, which then gives a structure of a 2D array. If you want to initialize the 2D array with a specific value, you need to pass it in the initializer. Consider this example:

   ```
   Array<IntArray>(10,{IntArray(10,{-1})})
   ```

3. The preceding 2D array will be initialized by all -1.

4. We can also use the arrayOf construct to create a 2D array by passing two 1D arrays:

   ```
   val even: IntArray = intArrayOf(2, 4, 6)
   val odd: IntArray = intArrayOf(1, 3, 5)
   ```

```
val lala: Array<IntArray> = arrayOf(even, odd)
lala.forEach {
    it.forEach {
        print(" ${it} ")
    }
    println()
}

//Output: 2  4  6
         1  3  5
```

There's more...

You can also create your own function by extending Kotlin's code. For example, creating a method, as follows:

```
inline fun <reified inside> array2d(sizeOuter: Int, sizeInner: Int,
noinline innerInit: (Int)->inside): Array<Array<inside>>
      = Array(sizeOuter) { Array<inside>(sizeInner, innerInit) }
```

This can enable you to create a 2D array quite easily, just by doing the following:

```
array2d(10,10,{0})
```

You can also create a list of lists in a similar fashion. Here's an example of a list of lists:

```
fun main(args: Array<String>) {
    val a= listOf(listOf(1,2,3), listOf(4,5,6), listOf(7,8,9))
    a.forEach {
        print(" ${it} ")
    }
}
```

This is its output:

```
[1, 2, 3] [4, 5, 6] [7, 8, 9]
```

How to skip the first "n" entries in Kotlin

In this recipe, we will learn how to drop entries in a collection. First, we will see how to drop the first *n* items, then we will see how to drop the last *n* items, and finally, we will see how to use a predicate while dropping the elements from the collection.

Getting ready

I'll be using IntelliJ IDEA for writing and running Kotlin code; you are free to use any IDE that can do the same task.

How to do it...

In the following steps, we will learn to skip the first *n* entries of a Kotlin list:

1. First, let's see how to drop the first *n* items in a collection. We will be using a list, but it also works with an array. Also, we will be using `kotlin.stdlib`, which contains the functions required in this recipe. The function to use here is `drop`:

```
fun main(args: Array<String>) {
    val list= listOf<Int>(1,2,3,4,5,6,7,8,9)
    var droppedList=list.drop(2)
    droppedList.forEach {
        print(" ${it} ")
    }
}

//Output: 3 4 5 6 7 8 9
```

2. To skip the last *n* items in the collection, you need to use the `dropLast` function:

```
fun main(args: Array<String>) {
    val list= listOf<Int>(1,2,3,4,5,6,7,8,9)
    var droppedList=list.dropLast(2)
    droppedList.forEach {
        print(" ${it} ")
    }
}

//Output:   1 2 3 4 5 6 7
```

3. This lambda function drops the item while the predicate returns true:

```
val list= listOf<Int>(1,2,3,4,5,6,7,8,9,1,2,3)
val droppedList=list.dropWhile { it<3 }
droppedList.forEach {
    print(" ${it} ")
}

//Output:   3 4 5 6 7 8 9 1 2 3
```

4. This function drops the items at the end while the condition is satisfied.

```
fun main(args: Array<String>) {
    val list= listOf<Int&gt;(1,2,3,4,5,6,7,8,9,3,1,2)
    val droppedList=list.dropLastWhile { it<3 }
    droppedList.forEach {
        print(" ${it} ")
    }
}

//Output:   1 2 3 4 5 6 7 8 9 3
```

How it works...

The `drop` function returns a new list by skipping the first n items. Internally, it just uses the usual `for` loops and performs some checks on whether the input is an array or a list.

7
Handling File Operations in Kotlin

The following recipes will be covered in this chapter:

- Reading from files using InputReader

- Reading all lines in a file using InputReader

- Reading line by line using InputReader

- Reading from files using BufferedReader

- Reading all lines in a file using BufferedReader

- Reading line by line using BufferedReader

- Reading string and JSON over network

Introduction

Kotlin I/O is used for input and output processing. Kotlin provides the `kotlin.io` API for working with files and streams. Some of the functions used in `kotlin.io` are extensions of `java.io` classes. All in all, using `kotlin.io` for reading from and writing to files and streams is very easy.

Reading from files using InputReader

`Kotlin.io` provides a nice, clean API for reading from and writing to files. One way of doing this is by using `InputReader`. We will see how to do that in this recipe.

Getting ready

You need to install a preferred development environment that compiles and runs Kotlin. You can also use the command line for this purpose, for which you need the Kotlin compiler installed along with JDK. You can also use IntelliJ IDEA for the development environment.

How to do it...

There are a lot of ways to go about reading from a file, but it is very important to understand the motivation behind them so as to be able to use the correct one for our purpose:

1. First, we will try to get the `InputStream` of the file and use the reader to read the contents:

```
import java.io.File
import java.io.InputStream

fun main(args: Array<String>) {
    val inputStream: InputStream = File("lorem.txt").inputStream()
    val inputString = inputStream.reader().use { it.readText() }
    println(inputString)
}
```

2. In the preceding code block, `lorem.txt` is simply a file that we want to read. The file is located in the same folder as our code source file. If we need to read a file located in a different folder, it looks similar to the following:

```
File("/path/to/file/lorem.txt")
```

3. This piece of code simply takes all the text in the file and prints it on the console.

4. Another way of reading file contents is by directly creating a reader of the file, like we do in this code:

```
import java.io.File

fun main(args: Array<String>) {
  val inputString = File("lorem.txt").reader().use { it.readText()
}
  println(inputString)
}
```

5. The output of both preceding code blocks will simply be the text in the file as it is. In our case, it was as follows:

```
Lorem ipsum dolor sit amet, consectetur adipiscing elit.
Nunc consequat eleifend mauris, eget congue ipsum consectetur id.
Proin hendrerit felis metus, vitae suscipit mi tempus facilisis.
Proin ut leo tellus. Donec nec lacus vel ante venenatis porttitor
et sit amet purus.
Sed tincidunt turpis ac metus pharetra dapibus.
Integer sed auctor tellus. Morbi a metus luctus, viverra enim vel,
imperdiet est.
Curabitur purus massa, hendrerit id ligula et, finibus elementum
purus.
In ut consectetur lacus.
Suspendisse non mauris eget dolor faucibus pharetra quis sed
turpis.
Vivamus eget lectus vel mi faucibus dignissim.
Class aptent taciti sociosqu ad litora torquent per conubia nostra,
per inceptos himenaeos.
Ut vitae velit non nunc consectetur imperdiet.
Nunc feugiat diam tellus, in pellentesque nisl dapibus quis.
Proin luctus sapien ac ante tempor, eget mollis odio aliquet.
```

6. Now, what if we want to read the file line by line because we want to do some processing on each line? In that case, we use the useLines() method in place of use().

7. Check out the following example, where we get an input stream from the file and use the useLines() method to get each line one after the other:

```
import java.io.File
import java.io.InputStream

fun main(args: Array<String>) {
```

```
        val listOfLines = mutableListOf<String>()
        val inputStream: InputStream = File("lorem.txt").inputStream()
        inputStream.reader().useLines { lines -> lines.forEach {
listOfLines.add(it)} }
        listOfLines.forEach{println("$ " + it)}
    }
```

8. Alternatively, if we wish to use a reader directly on the file, we do this:

```
    import java.io.File

    fun main(args: Array<String>) {
        val listOfLines = mutableListOf<String>()

        File("lorem.txt").reader().useLines { lines -> lines.forEach {
listOfLines.add(it)} }
        listOfLines.forEach{println("$ " + it)}
    }
```

The output, in this case, will be the following:

```
$ Lorem ipsum dolor sit amet, consectetur adipiscing elit.
$ Nunc consequat eleifend mauris, eget congue ipsum consectetur id.
$ Proin hendrerit felis metus, vitae suscipit mi tempus facilisis.
$ Proin ut leo tellus. Donec nec lacus vel ante venenatis porttitor et sit
amet purus.
$ Sed tincidunt turpis ac metus pharetra dapibus.
$ Integer sed auctor tellus. Morbi a metus luctus, viverra enim vel,
imperdiet est.
$ Curabitur purus massa, hendrerit id ligula et, finibus elementum purus.
$ In ut consectetur lacus.
$ Suspendisse non mauris eget dolor faucibus pharetra quis sed turpis.
$ Vivamus eget lectus vel mi faucibus dignissim.
$ Class aptent taciti sociosqu ad litora torquent per conubia nostra, per
inceptos himenaeos.
$ Ut vitae velit non nunc consectetur imperdiet.
$ Nunc feugiat diam tellus, in pellentesque nisl dapibus quis.
$ Proin luctus sapien ac ante tempor, eget mollis odio aliquet.
```

How it works...

Did you note that we used the `use()` and `useLines()` methods for reading the file? The call to the `Closeable.use()` function will automatically close the input at the end of the lambda's execution. Now, we can of course use `Reader.readText()`, but that does not close the stream after execution. There are other methods apart from `use()`, such as `Reader.readText()` and so on, that can be used to read the contents of a stream or file. The decision to use any method is based on whether we want the stream to be closed on its own after execution, or we want to handle closing the resources, and whether or not we want to read from a stream or directly from the file.

There's more...

`BufferedReader` reads a couple of characters at a time from the input stream and stores them in the buffer. That's why it is called `BufferedReader`. On the other hand, `InputReader` reads only one character from the input stream and the remaining characters still remain in the stream. There is no buffer in this case. This is why `BufferedReader` is fast, as it maintains a buffer, and retrieving data from the buffer is always quicker compared to retrieving data from disk.

Reading all lines in a file using InputReader

We can use `InputReader` to read all the lines in a file in one go. In this recipe, we will learn how to do that.

Getting ready

You need to install a preferred development environment that compiles and runs Kotlin. You can also use the command line for this purpose, for which you need the Kotlin compiler installed along with JDK. You can also use IntelliJ IDEA for the development environment.

How to do it...

Let's follow these steps to understand how to read a file using the `InputReader` class:

1. There are two ways to read a file, one of which is to attach an input stream to the file. Let's see how we can do that and use `InputReader` to read its contents:

```
import java.io.File
import java.io.InputStream

fun main(args: Array<String>) {
    val inputStream: InputStream =
File("example2.txt").inputStream()
    val inputString = inputStream.reader().use { it.readText() }
    println(inputString)
}
```

2. The other way is without getting a stream and directly reading the contents of the file, such as in the following example:

```
import java.io.File
fun main(args: Array<String>) {
    val inputString = File("example2.txt").reader().use {
it.readText() }
    println(inputString)
}
```

The output, in this case, is simply the contents of the file as it is:

```
A panoramic view of Lower Manhattan as seen at dusk from Jersey City, New
Jersey, in November 2014. Manhattan is the most densely populated borough
of New York City. It is the city's economic and administrative center, and
a major global cultural, financial, media, and entertainment center.
The second paragraph of this file is small.
```

We used the `use()` method because it closes the stream after execution.

How it works...

Attaching `inputStream` to a file returns the file contents as a stream of bytes. We can use a reader on the stream returned, or we can use the reader directly on the file. The `read()` method of `inputStream` reads the next byte in the stream. The `readText()` method returns the entire contents of the file as a string using UTF-8 or specified charset.

This `readText()` method is not recommended for huge files. It has an internal limitation of 2 GB file size. In case of a huge file, we read it byte by byte from the stream.

Reading line by line using InputReader

Sometimes we need to read the contents of a file line by line and process it. This is easily done by reading a file line by line using the `InputReader`. Let's see how.

Getting ready

You need to install a preferred development environment that compiles and runs Kotlin. You can also use the command line for this purpose, for which you need the Kotlin compiler installed along with JDK. You can also use IntelliJ IDEA for the development environment.

How to do it...

In the following steps, we will learn how to make use of the `InputReader` class to read the text line by line:

1. Let's start with attaching the `InputStream` to the file and going line by line on the contents, like this:

```
import java.io.File
import java.io.InputStream

fun main(args: Array<String>) {
    val listOfLines = mutableListOf<String>()
    val inputStream: InputStream =
File("example2.txt").inputStream()
    inputStream.reader().useLines { lines -> lines.forEach {
listOfLines.add(it)} }
    listOfLines.forEach{println("* " + it)}
}
```

2. Each line is appended with * in this case. Here's how the output looks:

```
* A panoramic view of Lower Manhattan as seen at dusk from Jersey
City, New Jersey, in November 2014. Manhattan is the most densely
populated borough of New York City. It is the city's economic and
```

administrative center, and a major global cultural, financial,
media, and entertainment center.
* The second paragraph of this file is small.

3. We can directly attach the reader to the file and read it line by line. The following code does just that:

```
import java.io.File

fun main(args: Array<String>) {
    val listOfLines = mutableListOf<String>()

    File("example2.txt").reader().useLines { lines -> lines.forEach
{ listOfLines.add(it)} }
    listOfLines.forEach{println("* " + it)}
}
```

How it works...

The `useLines()` method provides us an iterable over all the lines of the file or stream and then does something with each line, which is a string. We add all the modified strings to a list and print them out.

Reading from files using BufferedReader

`BufferedReader` stores some characters as it reads into the buffer. This makes the reading faster and hence more efficient. In this recipe, we will understand how to use the `BufferedReader` to read the contents of a file.

Getting ready

You need to install a preferred development environment that compiles and runs Kotlin. You can also use the command line for this purpose, for which you need the Kotlin compiler installed along with JDK. You can also use IntelliJ IDEA for the development environment.

How to do it...

Follow the mentioned steps to learn more about the working of the `BufferedReader` class:

1. We can directly attach a `BufferedReader` to the file and read the contents of the whole file, as in the following code:

```
import java.io.File
import java.io.InputStream

fun main(args: Array<String>) {
    val inputString = File("lorem.txt").bufferedReader().use {
it.readText() }
    println(inputString)
}
```

2. We can also go line by line on the contents that we need so as to be able to process each line individually. In the following code, we go line by line and add a character at the beginning and the length of the string after the character:

```
import java.io.File
import java.io.InputStream

fun main(args: Array<String>) {
    val listOfLines = mutableListOf<String>()
    File("lorem.txt").bufferedReader().useLines {
        lines -> lines.forEach {
            var x = "> (" + it.length + ") " + it;
            listOfLines.add(x)
        }
    }
    listOfLines.forEach{println(it)}
}
```

3. In the preceding code blocks, we are directly attaching the reader to the file. However, there are cases when we need to take a stream of data. In that scenario, we can get an Input Stream from the file that we want to read and then attach a `BufferedReader` to it.

4. In the following code, we are trying to read line by line from the file input stream using a `BufferedReader`:

```
import java.io.File
import java.io.InputStream

fun main(args: Array<String>) {
```

```kotlin
        val listOfLines = mutableListOf<String>()
        val inputStream: InputStream = File("lorem.txt").inputStream()
        inputStream.bufferedReader().useLines {
            lines -> lines.forEach {
                var x = "> (" + it.length + ") " + it;
                listOfLines.add(x)
            }
        }
        listOfLines.forEach{println(it)}
    }
```

5. Here's the output when we try to read all the contents of the file in one go:

    ```
    Lorem ipsum dolor sit amet, consectetur adipiscing elit.
    Nunc consequat eleifend mauris, eget congue ipsum consectetur id.
    Proin hendrerit felis metus, vitae suscipit mi tempus facilisis.
    Proin ut leo tellus. Donec nec lacus vel ante venenatis porttitor
    et sit amet purus.
    Sed tincidunt turpis ac metus pharetra dapibus.
    Integer sed auctor tellus. Morbi a metus luctus, viverra enim vel,
    imperdiet est.
    Curabitur purus massa, hendrerit id ligula et, finibus elementum
    purus.
    In ut consectetur lacus.
    Suspendisse non mauris eget dolor faucibus pharetra quis sed
    turpis.
    Vivamus eget lectus vel mi faucibus dignissim.
    Class aptent taciti sociosqu ad litora torquent per conubia nostra,
    per inceptos himenaeos.
    Ut vitae velit non nunc consectetur imperdiet.
    Nunc feugiat diam tellus, in pellentesque nisl dapibus quis.
    Proin luctus sapien ac ante tempor, eget mollis odio aliquet.
    ```

6. The output resembles the file, ignoring the `charset`. We can also specify the desired `charset`, such as we do in the following code, if needed:

    ```
    bufferedReader(charset).use { it.readText() }
    ```

7. When we go line by line using either of the preceding examples, we get the following output:

    ```
    > (56) Lorem ipsum dolor sit amet, consectetur adipiscing elit.
    > (65) Nunc consequat eleifend mauris, eget congue ipsum
    consectetur id.
    > (64) Proin hendrerit felis metus, vitae suscipit mi tempus
    facilisis.
    > (84) Proin ut leo tellus. Donec nec lacus vel ante venenatis
    ```

```
porttitor et sit amet purus.
> (47) Sed tincidunt turpis ac metus pharetra dapibus.
> (81) Integer sed auctor tellus. Morbi a metus luctus, viverra
enim vel, imperdiet est.
> (71) Curabitur purus massa, hendrerit id ligula et, finibus
elementum purus.
> (24) In ut consectetur lacus.
> (68) Suspendisse non mauris eget dolor faucibus pharetra quis sed
turpis.
> (46) Vivamus eget lectus vel mi faucibus dignissim.
> (91) Class aptent taciti sociosqu ad litora torquent per conubia
nostra, per inceptos himenaeos.
> (46) Ut vitae velit non nunc consectetur imperdiet.
> (60) Nunc feugiat diam tellus, in pellentesque nisl dapibus quis.
> (61) Proin luctus sapien ac ante tempor, eget mollis odio
aliquet.
```

How it works...

Using `InputStream` helps us get a stream of the file we wish to read. We can also read from the file directly though. In either case, the `BufferedReader` keeps presaving some data that it is reading in its buffer for faster operation, which increases the overall efficiency of the read operation, as compared to when using `InputReader`.

We use the `use()` and/or `useLines()` method in place of `Reader.readText()` and so on so that it automatically closes the input stream at the end of execution, which is a much cleaner and more responsible way of handling I/O of files. However, if needed, one can use a method such as `Reader.readText()` when they want to handle opening and closing the stream on their own.

Reading all lines in a file using BufferedReader

`BufferedReader` can be used to read contents of a file or an input stream. It presaves some contents it reads, so the read operation is faster. In this recipe, we will learn how to read all the contents of a file in one go using `BufferedReader`.

Getting ready

You need to install a preferred development environment that compiles and runs Kotlin. You can also use the command line for this purpose, for which you need the Kotlin compiler installed along with JDK. You can also use IntelliJ IDEA for the development environment.

How to do it...

In the following steps, we will learn how to use `BufferedReader` to read all lines of a file:

1. Let's start with getting the `InputStream` of our file and use the `BufferedReader` on it to read the contents of the file in one go:

```
import java.io.File
import java.io.InputStream
fun main(args: Array<String>) {
    val inputStream: InputStream = File("lorem.txt").inputStream()
    val inputString = inputStream.bufferedReader().use {
it.readText() }
    println(inputString)
}
```

2. The output, in this case, will be the exact same as the file, depending on the charset of course, if we use it. Here's an example where we use another charset:

```
import java.io.File
import java.io.InputStream
fun main(args: Array<String>) {
    val inputStream: InputStream = File("lorem.txt").inputStream()
    val inputString =
inputStream.bufferedReader(Charsets.ISO_8859_1).use {
it.readText() }
    println(inputString)
}
```

3. Now, let's quickly see the code example without getting the `inputStream` on this file:

```
import java.io.File
import java.io.InputStream
fun main(args: Array<String>) {
    val inputString = File("lorem.txt").bufferedReader().use {
it.readText() }
```

```
    println(inputString)
}
```

4. Although you might have guessed the output, here's the output anyway:

```
Lorem ipsum dolor sit amet, consectetur adipiscing elit.
Nunc consequat eleifend mauris, eget congue ipsum consectetur id.
Proin hendrerit felis metus, vitae suscipit mi tempus facilisis.
Proin ut leo tellus. Donec nec lacus vel ante venenatis porttitor
et sit amet purus.
Sed tincidunt turpis ac metus pharetra dapibus.
Integer sed auctor tellus. Morbi a metus luctus, viverra enim vel,
imperdiet est.
Curabitur purus massa, hendrerit id ligula et, finibus elementum
purus.
In ut consectetur lacus.
Suspendisse non mauris eget dolor faucibus pharetra quis sed
turpis.
Vivamus eget lectus vel mi faucibus dignissim.
Class aptent taciti sociosqu ad litora torquent per conubia nostra,
per inceptos himenaeos.
Ut vitae velit non nunc consectetur imperdiet.
Nunc feugiat diam tellus, in pellentesque nisl dapibus quis.
Proin luctus sapien ac ante tempor, eget mollis odio aliquet.
```

How it works...

BufferedReader, as it reads, stores some characters in a buffer, hence the read operations are faster. We can directly attach the BufferedReader to the file or the stream and read from it.

The use() method ensures that the file or stream is closed after execution completes.

Reading line by line using bufferedReader

In this recipe, we will understand how to use the bufferedReader to read the contents of a file line by line.

Getting ready

You need to install a preferred development environment that compiles and runs Kotlin. You can also use the command line for this purpose, for which you need the Kotlin compiler installed along with JDK. You can also use IntelliJ IDEA for the development environment.

How to do it...

In the given steps, we will learn how to use BufferedReader to read a file line by line:

1. Let's start with getting the InputStream of our file and use the BufferedReader on it to read the contents of the file line by line:

```kotlin
import java.io.File
import java.io.InputStream
fun main(args: Array<String>) {
    val listOfLines = mutableListOf<String>()
    val inputStream: InputStream = File("lorem.txt").inputStream()
    inputStream.bufferedReader().useLines {
        lines -> lines.forEach {
            var x = "# (" + it.length + ") " + it.substring(0,8);
            listOfLines.add(x)
        }
    }
    listOfLines.forEach{println(it)}
}
```

The output, in this case, is as follows:

```
# (56) Lorem ip
# (65) Nunc con
# (64) Proin he
# (84) Proin ut
# (47) Sed tinc
# (81) Integer
# (71) Curabitu
# (24) In ut co
# (68) Suspendi
# (46) Vivamus
# (91) Class ap
# (46) Ut vitae
# (60) Nunc feu
# (61) Proin lu
```

2. The output depends on the charset we use if we use one. This is a code example with a charset:

```
import java.io.File
import java.io.InputStream
fun main(args: Array<String>) {
    val listOfLines = mutableListOf<String>()
    val inputStream: InputStream = File("lorem.txt").inputStream()
    inputStream.bufferedReader(Charsets.US_ASCII).useLines {
        lines -> lines.forEach {
            var x = "# (" + it.length + ") " + it.substring(0,8);
            listOfLines.add(x)
        }
    }
    listOfLines.forEach{println(it)}
}
```

3. Now, let's go through a code example where we read directly from a file:

```
import java.io.File
import java.io.InputStream
fun main(args: Array<String>) {
    val listOfLines = mutableListOf<String>()
    File("lorem.txt").bufferedReader().useLines {
        lines -> lines.forEach {
            var x = "# (" + it.length + ") " + it.substring(0,8);
            listOfLines.add(x)
        }
    }
    listOfLines.forEach{println(it)}
}
```

How it works...

`BufferedReader` provides us with a lot of methods that we can use to read contents of the file or input stream line by line. Using `useLines()`, we get a sequence of lines that we can then iterate on using `forEach`. A user may terminate the iteration loop, so the caller needs to close the `BufferedReader`, which is what `useLines()` does. We can only iterate over the sequence returned once.

The syntax for `useLines()` is this:

```
inline fun <T> File.useLines(
    charset: Charset = Charsets.UTF_8,
    block: (Sequence<String>) -> T
): T
```

There's more...

We can also use other methods like `readLine()` for this purpose. This code is an example of that:

```
import java.io.File
import java.io.InputStream
fun main(args: Array<String>) {
    val listOfLines = mutableListOf<String>()
    val reader = File("lorem.txt").bufferedReader()
    while(true) {
        var line = reader.readLine()
        if(line == null) break
        listOfLines.add("> "+line)
    }
    listOfLines.forEach{println(it)}
}
```

The great thing about using the `useLines()` method is that it closes the stream post-execution. Also, the code in the preceding examples was a more idiomatic and clean way of doing the same thing.

Another method provided by Kotlin that returns a sequence of lines is `lineSequence()`, but it does not close the `BufferedReader` after execution, which is why it's good to use `useLines()`.

In the end, it depends on the scenario in which the code is to be used.

Reading string and JSON over network

Networking is an essential component of apps. Most of the apps we use are connected to the internet and involve reading/writing data over the internet. In this recipe, we will learn how to perform network requests in Kotlin. Although you can also use a third-party library such as Retrofit, Volley and such, understanding how it's done in Kotlin is worthwhile. So let's get started!

Getting ready

We will be working with Android code, so I'll be using Android Studio. It's also required to include the anko-commons library as we will be using its methods in our code.

How to do it...

Let's follow these steps to understand how to make a network request in Kotlin:

1. Making a network request in Kotlin is straightforward with simple syntax. Here's how you would read data over the internet in Kotlin:

```
val response = URL("<api_request>").readText()
```

2. That's it! Remember that this is equivalent to Java code when making a network request:

```
// 1. Declare a URL Connection
URL url = new URL("http://www.google.com");
HttpURLConnection conn = (HttpURLConnection) url.openConnection();
// 2. Open InputStream to connection
conn.connect();
InputStream in = conn.getInputStream();
// 3. Download and decode the string response using builder
StringBuilder stringBuilder = new StringBuilder();
BufferedReader reader = new BufferedReader(new
InputStreamReader(in));
String line;
while ((line = reader.readLine()) != null) {
    stringBuilder.append(line);
}
```

3. However, of course, if you try it on the main thread, you will get the
   NetworkOnMainThreadException exception. To get away with this, we need to
   make the network call in the background. One way to do this is by using
   an Async task. An Async task was a pain to implement in Java, but we can do it
   quite easily using Anko (a library for Kotlin). This is how you would create a
   background task in Kotlin using Anko:

```
doAsync{
    val response=URL("<network_url>").readText()
    uiThread{
        // Here you would do UI operation
        toast(" ... ")
```

```
        }
    }
```

4. In Java's `Async` task implementation, the `Async` task could be fired even if the activity was being destroyed. This resulted in defensive programming where you had to make checks on whether the UI was still present to do UI operations. However, Anko's implementation of background task takes care of it, and it won't fire the task if the activity is dying.

How it works...

The `doAsync` returns a Java `Future`. In simple words, a `Future` is a proxy or a wrapper around an object that is not yet there. When the asynchronous operation is done, you can extract it. If you want to avoid working with `Future`, `doAsync` has a different construct that accepts an `ExecutorService`:

```
val executor = Executors.newScheduledThreadPool(5)

doAsync(executorService = executor){
    val result = URL("https://httpbin.org/get").readText()
    uiThread {
        toast(result)
    }
}
```

As we discussed, the `uiThread` block isn't executed if the activity is closing. The reason is that it doesn't hold a context instance, only a weak reference. So even if the block isn't finished, the context will not leak.

8
Anko Commons and Extension Function

The following recipes will be covered in this chapter:

- Setting up Anko with Gradle

- Extending Android framework using extension function

- Using extensions as properties

- Using intents with Anko

- Making a call intent using Anko

- Sending a text intent using Anko

- Browsing the web browser using Anko

- Sharing some text using intents in Anko

- Sending an email using Anko

- Creating Android dialogs with Anko

- Showing an alert dialog with a list of text items

- Using Anko in Views

- Logging using Anko

- Handling dimensions with Anko

- Version checking in Android

Introduction

Anko is a Kotlin library that was developed to make the Android development experience better. Kotlin itself makes Android development easier, and Anko is a cherry on top of it. With helpers for almost all common Android functionalities, Anko drastically reduces the amount of code you write and makes Android development fun.

Anko consists of several parts:

- **Anko Commons**: It consists of helper methods for intents, dialogs, logging, and so on, which reduces the amount of code significantly.
- **Anko Layouts**: With this library, you don't have to stick to conventional XML to create visual interfaces. Anko layout is a fast and type-safe way to write dynamic Android layouts.
- **Anko SQLite**: This is a query DSL and parser collection for Android SQLite, which makes working with underlying SQLite database substantially easy.
- **Anko Coroutines**: Coroutines are a great way to do asynchronous programming. Anko coroutines provide utilities based on the `kotlinx.coroutines` (`https://github.com/Kotlin/kotlinx.coroutines`) library.

In this chapter, we will learn how to use Anko for Android development. So let's get started!

Setting up Anko with Gradle

We will begin by setting up the Anko library in our project. We will be using Gradle to handle the dependencies of the project.

Getting ready

I'll be using Android Studio to write code. You can also find the source code in the **1-setting-up-anko-with-gradle** branch of repository at `https://gitlab.com/aanandshekharroy/Anko-examples`.

How to do it...

Follow these steps to add Anko to your project using the Gradle build system:

1. The easiest way to set up Anko with Gradle is do it by adding the following lines in your `build.gradle` file:

```
compile "org.jetbrains.anko:anko:$anko_version"
```

2. You can replace `$anko_version` with the latest version of Anko, which is 0.10.1 when this book was written.

3. The preceding compile statement will add all available features (including Commons, Layouts, SQLite) into your project at once. If you don't want that, and would prefer adding them separately as needed, here are the compile statements:

- `anko-commons`: This library contains a lot of helpers for Android SDK for Intents, Dialogs and Toasts, Logging, and Resource and Dimension:

```
compile "org.jetbrains.anko:anko-commons:$anko_version"
```

- **Anko Layouts**: Anko Layouts is a DSL for writing dynamic Android layouts:

```
compile "org.jetbrains.anko:anko-sdk25:$anko_version" //
sdk15,19,21,23 are also available
compile "org.jetbrains.anko:anko-appcompat-v7:$anko_version"
```

- `anko-sqlite`: This provides helpers for working with SQLite database:

```
compile "org.jetbrains.anko:anko-sqlite:$anko_version"
```

- `anko-coroutines`: This library makes it easier to work with Kotlin coroutines:

```
compile "org.jetbrains.anko:anko-coroutines:$anko_version"
```

Extending Android framework using extension function

The heading of this recipe might seem very confusing to you as you might be thinking "How can I extend the super complex Android framework? And moreover, why should I?" We will be dealing with all the "what, why, and hows" of extension functions in this recipe. Extension functions are one of the greatest features of Kotlin. So let's dive into it.

Getting ready

I'll be using Android Studio for coding. We will be creating extension functions for Android SDK classes.

How to do it...

To begin with, let's see a very simple example:

1. We will create a very simple class `Student`, and we will create an extension function for it:

```
class Student(val age:Int)
```

2. Now, we would want to create an `isAgeGreaterThan20` function, which will return `true` if the age is greater than 20, or else, will return `false`. Now suppose there's a restriction that we can't touch the Student class, what will you do?

3. In those scenarios, extension functions come in handy when you want to extend the functionality of a class. If you try to call the method, you will be shown an error, as follows:

```
fun main(args: Array<String>) {
    val studentA=Student( age: 25)
    println(studentA.isAgeGreaterThan20())
}
class Student(val age:Int)
```
    💡 Create member function 'Student.isAgeGreaterThan20'
    💡 Rename reference
    💡 Create extension function 'Student.isAgeGreaterThan20'

4. You then need to select the `Create extension function ....` option to create an extension function for it. When you select that option, you'll again be given two options, asking which object you want to create the extension function.

```
fun main(args: Array<String>) {
    val studentA=Student( age: 25)
    println(studentA.isAgeGreaterThan20())
}

class Student(val age:Int)
```
    Choose target class or interface
    Student   KotlinCookbook
    Any   KotlinJavaRuntime (kotlin-stdlib.jar)

5. Since we want to create the function for the `Student` class, we will select the **Student** option from the dropdown. On selecting that, the IDE will autogenerate the method body. I've modified the return type to return a boolean:

```
private fun Student.isAgeGreaterThan20(): Boolean {
}
```

6. Then, we can do the operation inside the method block. Here's how our method looks:

```
private fun Student.isAgeGreaterThan20(): Boolean {
    return this.age>20
}
```

7. Note that since we are calling the method on a student object, we can access it using the `this` keyword, though you can skip the `this` keyword in this case, as we aren't dealing with other objects of the same type in this method.

8. Now, we can call it just like you would call a normal method:

```
fun main(args: Array<String>) {
    val studentA=Student(25)
    println(studentA.isAgeGreaterThan20())
}
>
//Output: true
```

9. Now, let's see an example related to Android. If you've used any third-party library like `Picasso` or `Glide`, you might remember setting images in `ImageView` like this:

```
Picasso.with(context).from(url).into(imageView);
```

10. You can create an extension function for `ImageView` named `loadImage` and then call that function in your application. Of course, `loadImage` is not a function provided by the `ImageView` class, so you need to create an extension function for this purpose. We will call the method on the `imageView` object and pass a `url`:

```
imageView.loadImage(url)
private fun ImageView.loadImage(url: String) {
    Picasso.with(this.context).load(url).into(this)
}
```

11. Note that in the `loadImage` function, we are referring `this` to the `ImageView` object on which the function was called.

How it works...

Extension functions are resolved statically, which means they are normal static methods and have no connection to the class they are extending (that's why we are able to extend classes to which we don't have access to modify), other than taking an instance of this class as a parameter.

If you decompile the Kotlin bytecode, you will see the code converted to Java:

```
private static final boolean isAgeGreaterThan20(@NotNull Student $receiver)
{
    return $receiver.getAge() > 20;
}
```

As you can see, it is just a static method and takes the class as a parameter.

There's more...

Since extension functions come in handy, you might be tempted to use them a lot. However, with great power comes great responsibility. Since they are resolved statically, you shouldn't be using them everywhere, because static functions are difficult to test. Using it irresponsibly means your code will be less testable and hence less maintainable.

Using extensions as properties

In the last recipe, we learned about extension functions. In this recipe, we will learn about extension properties. If you feel the need for one or more properties from the class, you can add them using the extension properties. In this recipe, we will learn how to use extension properties.

Getting ready

I'll be using Android Studio for coding purposes. Ensure that you have the latest version of Android Studio with Kotlin configured.

How to do it...

Let's see an example of an extension property now:

1. We will be using the example of shared preferences. You might be used to doing something like this to get hold of shared preferences:

```
PreferenceManager.getDefaultSharedPreferences(this)
```

2. You can create an extension property on the `Context` class with name preferences and access it as follows:

```
val Context.preferences: SharedPreferences
        get() = PreferenceManager
        .getDefaultSharedPreferences(this)
context.preferences.getInt("...")
```

How it works...

The extension functions didn't modify the class, and the same goes with extension properties; they don't add properties to the class itself and hence we don't have any backing field in their case. Also, since we don't have any backing field, we can't initialize it. The only way too deal with them is using custom getter and setter.

There's more...

Similar to extension properties, we can have companion object extensions, which means we can add methods to the companion object of a class that helps us access it in a static way. Let's look at an example. Suppose we have a `Student` class:

```
class Student(val age:Int){
    companion object{
    }
}
```

Let's now add an extension method to the companion object:

```
fun Student.Companion.sayHi(){
    println("Hi")
}
```

Now you can access it on the class without creating an instance of the class:

```
Student.sayHi()
```

Using intents with Anko

Intents are one of the most common components used in Android apps. They can be thought of as a messenger used to transfer messages between different Android components. For example, you send an intent when you need to start an activity, you send an intent when you need to start a service. To launch an activity in Android, you are first needed to create an intent and then pass it to the `startActivity` method. In the following example, we will try to launch an activity with some data and flags:

```
val intent = Intent(this, SomeActivity::class.java)
intent.putExtra("data", 5)
intent.setFlag(Intent.FLAG_ACTIVITY_SINGLE_TOP)
startActivity(intent)
```

Also, you can assume an extra line for all data that you pass with the intent.

Anko has a better way to achieve similar results. In this recipe, we will learn how to achieve this (launching intents) using Anko library.

Getting ready

I'll be using Android Studio for coding purpose. You need to include Anko library in your `build.gradle` file at app level. Just add these lines and you are good to go:

```
compile "org.jetbrains.anko:anko-commons:$anko_version"
```

How to do it...

Creating an intent in Anko is very simple. Let's check the following steps:

1. The functionality for which we wrote the preceding code can be achieved in just a few lines with Anko:

```
startActivity(intentFor<SomeActivity>("data" to 5).singleTop())
```

2. If you don't want to add the flag, it is much simpler:

```
startActivity<SomeActivity>("data" to 5)
```

3. Adding extra data doesn't require extra lines:

```
startActivity<SomeActivity>("data" to 5, "another_data" to 10)
```

How it works...

Let's see the implementation of the preceding methods in the source code:

```
inline fun <reified T: Any> Context.intentFor(vararg params: Pair<String,
Any?>)
```

The `intentFor` method takes `vararg` as a parameter, hence we are able to supply multiple data to it. This method calls `createIntent`, which actually creates an intent with supplied data, and it looks like this:

```
fun <T> createIntent(ctx: Context, clazz: Class<out T>, params: Array<out
Pair<String, Any?>>): Intent {
    val intent = Intent(ctx, clazz)
    if (params.isNotEmpty()) fillIntentArguments(intent, params)
    return intent
}
private fun fillIntentArguments(intent: Intent, params: Array<out
Pair<String, Any?>>) {
    params.forEach {
        val value = it.second
        when (value) {
            null -> intent.putExtra(it.first, null as Serializable?)
            is Int -> intent.putExtra(it.first, value)
            is Long -> intent.putExtra(it.first, value)
            is CharSequence -> intent.putExtra(it.first, value)
            is String -> intent.putExtra(it.first, value)
            is Float -> intent.putExtra(it.first, value)
            is Double -> intent.putExtra(it.first, value)
            is Char -> intent.putExtra(it.first, value)
            is Short -> intent.putExtra(it.first, value)
            is Boolean -> intent.putExtra(it.first, value)
            is Serializable -> intent.putExtra(it.first, value)
            is Bundle -> intent.putExtra(it.first, value)
            is Parcelable -> intent.putExtra(it.first, value)
            is Array<*> -> when {
```

```
                        value.isArrayOf<CharSequence>() ->
    intent.putExtra(it.first, value)
                        value.isArrayOf<String>() -> intent.putExtra(it.first,
    value)
                        value.isArrayOf<Parcelable>() -> intent.putExtra(it.first,
    value)
                        else -> throw AnkoException("Intent extra ${it.first} has
    wrong type ${value.javaClass.name}")
                    }
                is IntArray -> intent.putExtra(it.first, value)
                is LongArray -> intent.putExtra(it.first, value)
                is FloatArray -> intent.putExtra(it.first, value)
                is DoubleArray -> intent.putExtra(it.first, value)
                is CharArray -> intent.putExtra(it.first, value)
                is ShortArray -> intent.putExtra(it.first, value)
                is BooleanArray -> intent.putExtra(it.first, value)
                else -> throw AnkoException("Intent extra ${it.first} has wrong
    type ${value.javaClass.name}")
            }
        return@forEach
    }
}
```

As you can see, internally it creates the intent in the old fashion and calls
`fillIntentArguments`, which fills the intent with the data.

Making a call intent using Anko

In the last recipe, we learned how to create an intent using Anko library. In subsequent
recipes, we will see how to do common things like sending messages, calls, mails, and so on
using intents in Anko.

Getting ready

I'll be using Android Studio for coding purposes. You need to include Anko library in your
`build.gradle` file. Just add the following lines to your `build.gradle` file and you are
good to go:

```
compile "org.jetbrains.anko:anko-commons:$anko_version"
```

You can also clone the `gitlab.com/aanandshekharroy/Anko-examples` repository and
switch to the **3-intent-actions** branch to get the source code.

How to do it...

Let's follow the given steps to make a call using intents:

1. Anko provides wrappers around the most common actions that can be done using intents; one of them is making calls. For this purpose, Anko provides the `makeCall` function, which takes in the phone number you want to call:

   ```
   makeCall("+9195XXXXXXXX")
   ```

2. The `makeCall` function returns true if the action was successful, or returns false if it wasn't. One thing to note is that you need to add the `CALL_PHONE` permission in your manifest file:

   ```
   <uses-permission android:name="android.permission.CALL_PHONE"/>
   ```

How it works...

Let's see the source code of `makeCall function`:

```
fun Context.makeCall(number: String): Boolean {
    try {
        val intent = Intent(Intent.ACTION_CALL, Uri.parse("tel:$number"))
        startActivity(intent)
        return true
    } catch (e: Exception) {
        e.printStackTrace()
        return false
    }
}
```

Beneath the wrapper, it's doing the same old way Android SDK used to do, that is, using an implicit intent of action `Intent.ACTION_CALL`.

Sending a text intent using Anko

Anko provides a wrapper around the intent actions, which makes calling actions super easy. One of those actions is sending an SMS. In this recipe, we will see how to launch an intent that sends messages to a telephone number.

Getting ready

I'll be using Android Studio for coding purposes. You need to include Anko library in your `build.gradle` file. Just add the given lines and you are good to go:

```
compile "org.jetbrains.anko:anko-commons:$anko_version"
```

You can also clone the `gitlab.com/aanandshekharroy/Anko-examples` repository and switch to the **3-intent-actions** branch to get the source code.

How to do it...

Let's follow these steps to send an SMS using intents:

1. Anko provides the `sendSMS` method, which takes in two parameters—one of them is the phone number, and the other is the message:

```
sendSMS("+9195XXXXXX","Hi")
```

2. Calling this method will launch the messaging app, or will ask you which messaging app to launch if you have more than one type of that app and will prefill the message body. Calling this function requires you to add the following permission, without which it will throw a Security Exception:

```
<uses-permission android:name="android.permission.SEND_SMS"/>
```

How it works...

To understand how its working, let's dive into its implementation:

```
fun Context.sendSMS(number: String, text: String = ""): Boolean {
    try {
        val intent = Intent(Intent.ACTION_VIEW, Uri.parse("sms:$number"))
        intent.putExtra("sms_body", text)
        startActivity(intent)
        return true
    } catch (e: Exception) {
        e.printStackTrace()
        return false
    }
}
```

As you can see, it uses an implicit intent to launch the messaging app on your device. Since this function requires a context, if you are calling it from a fragment, you need to call it as `activity.sendSMS(..)`.

Browsing the web browser using Anko

In this recipe, we will talk about the Anko wrapper that will help us browse the website using a web browser. So let's get started.

Getting ready

I'll be using Android Studio for coding purpose. You need to include Anko library in your `build.gradle` file. Just add this line of code and you are good to go:

```
compile "org.jetbrains.anko:anko-commons:$anko_version"
```

You can also clone the `gitlab.com/aanandshekharroy/Anko-examples` repository and switch to the `3-intent-actions` branch to get the source code.

How to do it...

Now, let's see how to launch a browser using an intent.

Anko provides a `browse` function, which takes in the web address and launches the browser on your device. If you have multiple browsers, it will show you some options to select it. Here's an example:

```
browse("http://www.google.com")
```

The web address you put in the parameter needs to have `http://` or `https://` as the prefix, otherwise it will throw an `ActivityNotFound` exception.

How it works...

The `browse` function provided by Anko is just a syntactic sugar, beneath which we have the same old code that we used previously:

```
fun Context.browse(url: String, newTask: Boolean = false): Boolean {
    try {
        val intent = Intent(Intent.ACTION_VIEW)
        intent.data = Uri.parse(url)
        if (newTask) {
            intent.addFlags(Intent.FLAG_ACTIVITY_NEW_TASK)
        }
        startActivity(intent)
        return true
    } catch (e: ActivityNotFoundException) {
        e.printStackTrace()
        return false
    }
}
```

Calling this method returns true or false, based on whether the action was successful or not.

Sharing some text using intents in Anko

In this recipe, we will look at how to share text using Anko wrapper. Sharing text is a very common thing and Anko provides a wrapper for this action that is very easy to use. So let's get started!

Getting ready

I'll be using Android Studio for coding purposes. You need to include Anko library in your `build.gradle` file. Just add the given lines and you are good to go:

```
compile "org.jetbrains.anko:anko-commons:$anko_version"
```

You can also clone the repository at `gitlab.com/aanandshekharroy/Anko-examples` and switch to the `3-intent-actions` branch to get the source code.

How to do it...

In the following steps, we will see how to share text using an intent:

1. Anko provides a `share` method, which takes in a string parameter, that is, the text to share and an optional parameter subject. The subject parameter can be particularly useful to share text via an email app. After all, who sends a subject for a whatsapp message, right? Let's see its implementation:

```
share("Hey","Some subject")
```

2. Without the subject—this won't fill in the subject line in the mail:

```
share("Hey")
```

It's that simple!

How it works...

If you look at the implementation, you will find that Anko has provided just a syntactic sugar that greatly reduces your lines of code to achieve a similar thing:

```
fun Context.share(text: String, subject: String = ""): Boolean {
    try {
        val intent = Intent(android.content.Intent.ACTION_SEND)
        intent.type = "text/plain"
        intent.putExtra(android.content.Intent.EXTRA_SUBJECT, subject)
        intent.putExtra(android.content.Intent.EXTRA_TEXT, text)
        startActivity(Intent.createChooser(intent, null))
        return true
    } catch (e: ActivityNotFoundException) {
        e.printStackTrace()
        return false
    }
}
```

As you can see, the library has taken care of all the issues that might crop up and just provided a helper function to make things faster and more fun.

Sending an email using Anko

In this recipe, we will see how to send an email using Anko's wrapper. Sending an email is very useful as almost all apps provide a method of contact. So let's get started!

Getting ready

I'll be using Android Studio for coding purposes. You need to include Anko library in your `build.gradle` file. Just add these lines and you are good to go:

```
compile "org.jetbrains.anko:anko-commons:$anko_version"
```

You can also clone the repository at `gitlab.com/aanandshekharroy/Anko-examples` and switch to the `3-intent-actions` branch to get the source code.

How to do it...

We will use the `email` function provided by Anko library that takes three parameters, out of which only one is mandatory:

```
email("support@XXXXXX.com","Subject","Text")
```

You can remove the subject and text if you don't want prefilled text in the email.

How it works...

Let's take a look at its implementation:

```
fun Context.email(email: String, subject: String = "", text: String = ""):
Boolean {
    val intent = Intent(Intent.ACTION_SENDTO)
    intent.data = Uri.parse("mailto:")
    intent.putExtra(Intent.EXTRA_EMAIL, arrayOf(email))
    if (subject.isNotEmpty())
        intent.putExtra(Intent.EXTRA_SUBJECT, subject)
    if (text.isNotEmpty())
        intent.putExtra(Intent.EXTRA_TEXT, text)
    if (intent.resolveActivity(packageManager) != null) {
        startActivity(intent)
        return true
    }
```

```
        return false

    }
```

As you can see, it checks for the extra data, such as the subject and message body, and then launches the email application. The email function provided by Anko is just a convenient method that reduces your line of code and makes your code look good.

Creating Android dialogs with Anko

One of the really great features of Anko library is that it can help you create alert dialog quite easily and with much less code.

In this recipe, we will see how to create alert dialogs in Anko.

Getting ready

I'll be using Android Studio to write code. You also need to include the Anko library by adding the following lines to your `build.gradle` file:

```
compile "org.jetbrains.anko:anko:$anko_version"
```

You can find the source code in the **2-creating-dialogs-using-anko** branch of the `https://gitlab.com/aanandshekharroy/Anko-examples/` repository.

How to do it...

Let's follow the mentioned steps to create a dialog in Kotlin:

1. In the first example, we will try to create a simple alert box. To create it, you just need to follow this syntax:

```
alert("A simple alert","Alert") {

    }.show()
```

2. If you try to run it, you will see something like this:

3. There are certain situations where you want the user to perform some actions, and so Anko provides you the methods for it. Check out the following example:

```
alert("Would you like some action?","Alert") {
    yesButton {
        toast("Clicked on Yes")
    }
    noButton {
        toast("Clicked on No")
    }
    neutralPressed("Meh"){
        toast("Not interest")
    }
}.show()
```

4. This inflates the alert dialog, as illustrated:

5. You can also customize the text of `yesButton` and `noButton` by replacing them with `positiveButton` and `negativeButton`. Here's an example:

```
alert("Would you like some action?","Alert") {
    positiveButton("Hell Yeah!") {
        toast("Clicked on Yes")
    }
    negativeButton("No way!") {
        toast("Clicked on No")
    }
    neutralPressed("Meh?"){
        toast("Not interest")
    }
}.show()
```

If you run the preceding code, you will see a dialog appearing on your device, as follows:

6. Another common type of dialog used in Android development is a progress dialog. You can use Anko to create a progress dialog like this:

7. This kind of progress dialogs is good for showing the user how much progress is made. It also provides functions such as `incrementProgressBy`, by which you can increase the progress bar. To create such a progress dialog, you need to use it like this example:

```
val dialog = progressDialog(message = "Please wait a bit...", title
= "Fetching data")
dialog.show()
```

8. Perhaps you would like to have an indefinite progress bar, which looks like this:

To create an indeterminate progress dialog as in the preceding screenshot, simply add the following lines to your previous code:

```
indeterminateProgressDialog("This is an indeterminate progress
dialog").show()
```

Showing an alert dialog with a list of text items

In the previous recipe, we saw how to create different types of dialogs. In this recipe, we will see how to create an alert dialog with a list of text items, which looks as illustrated in the following screenshot:

Getting ready

I'll be using Android Studio to write code. You also need to include the Anko library by adding these lines to your `build.gradle` file:

```
compile "org.jetbrains.anko:anko:$anko_version"
```

How to do it...

Let's go through the given steps to create an alert dialog with a list of items.

Anko provides selectors for creating a dialog with a list of items. Selectors are very easy to use. You just need to provide the title of alert dialog, the list, and the lambda that will be executed when an option is selected. Here's an implementation of it:

```
val companies = listOf("Google", "Microsoft", "HP", "Apple")
selector("Where do you work?", companies, { dialogInterface, i ->
    toast("So you work at ${companies[i]}, right?")
})
```

That's all folks! It's really simple and concise. So, in the preceding example, if you click on the list item, you will see a toast with a message that says "So you work at Google, right?"

How it works...

Anko hides all the complexity and gives you an easy-to-use function to achieve complex things. Let's check out the implementation of the selector function:

```
fun Context.selector(
        title: CharSequence? = null,
        items: List<CharSequence>,
        onClick: (DialogInterface, Int) -> Unit
) {
    with(AndroidAlertBuilder(this)) {
        if (title != null) {
            this.title = title
        }
        items(items, onClick)
        show()
    }
}
```

As you can see, beneath the surface, it's just like the old way, but Anko provides syntactic sugar, which helps us achieve the same thing with less code.

Using Anko in Views

Anko makes handling views and creating layouts extremely fast and easy. Using Anko, we can write clean code that is easy to read and write. In this recipe, we will learn how Anko can be used when dealing with views in Android.

Getting ready

I'll be using Android Studio 3 to write code. You can get started by creating a new project in Kotlin with a blank activity in Android Studio 3+ as we won't be using any code from other recipes. You also need an intermediate understanding of Android development. Ensure that you have added Anko dependencies to your project by adding the following lines to your app level `build.gradle` file and syncing the project:

```
compile "org.jetbrains.anko:anko:$anko_version"
```

Here, `$anko_version` is the latest version of Anko out there.

How to do it...

Anko makes some common Android development stuff extremely easy, such as toasts, snackbars, and dialogs. Usually, showing these views takes a lot of code. Let's see how it is just a matter of a few lines of easy code with Anko:

Alert dialog: A popup that appears on top of your view, commonly used for alerts:

- To show an alert, we use the following syntax (DSL syntax):

```
alert("Hi, I'm Moss", "This, Jen, is the internet") {
    yesButton { toast("Oh...") }
    noButton {toast("Well...") }
}.show()
```

- Consider that we use a dialog from the `Appcompat` dialog factory:

```
alert(Appcompat, "Hello, Jen.").show()
```

- We can also show progress dialogs and indeterminate progress dialogs:

```
val dialog = progressDialog(message = "Please stand by", title =
"Fetching data")

indeterminateProgressDialog("You just have to wait indefinitely
Jen.").show()
```

Toasts: They can be used to display information for a short period of time:

- To show a toast, we can make use of one of the following syntaxes based on the need of the situation:

```
toast("Hi! I'm Roy")
toast(R.string.meet_roy)
longToast("We have been together for a long time.")
```

Snackbar: Snackbars are just like Toast messages, except that they provide action to interact with:

- There are different ways to show snackbars, based on whether you use string or string resource and what is the duration of timeout of the snackbar and whether you need an action button or not. For showing snackbars, you need a reference to the parent view in which you wish to show a snackbar. In the case of XML, you can find a method by Anko to get the view from its ID or in the case of DSL, you can directly use the variable in which you have stored the parent view. Here are some syntaxes that we can use:

```
snackbar(rootView, "Hi! I'm Jen")
snackbar(rootView, R.string.go_away_jen)
longSnackbar(rootView, "I'm going to be here for a long time")
snackbar(rootView, "What do you want?", "Click me") { doSomething()
}
```

Anko makes it easier to define layouts and handle already created layouts (in XML).

Creating a layout in DSL:

- Creating a layout in DSL is very simple, and we can put it directly in the onCreate() method of activity, as in the following code:

```
lateinit var rootView: View
lateinit var btn: Button
lateinit var editText1: EditText
lateinit var editText2: EditText
```

```
override fun onCreate(savedInstanceState: Bundle?) {
    super.onCreate(savedInstanceState)
    rootView = verticalLayout {
        padding = dip(20)
        editText1 = editText {
            hint = "What's your name?"
        }

        editText2 = editText {
            hint = "What's your message?"
        }
        btn = button("Click me") {
            onClick {
                toast( "Hey! Here is a toast for you.")
            }
        }
    }
}
```

- Alternatively, we can put it in an external class that implements the AnkoComponent interface:

```
class MainActivity : AppCompatActivity() {

    override fun onCreate(savedInstanceState: Bundle?) {
        super.onCreate(savedInstanceState)
        MainActivityUI().setContentView(this)
    }

    class MainActivityUI : AnkoComponent<MainActivity> {
        override fun createView(ui: AnkoContext<MainActivity>) =
with(ui) {
            verticalLayout {
                padding = dip(20)

                editText {
                    hint = "What's your name?"
                }

                editText {
                    hint = "What's your message?"
                }

                button("Click me") {
                    onClick {
                        toast( "Hey! Here is a toast for you.")
                    }
```

```
                    }
                }
            }
        }
    }
```

- Also, this is how our layout looks:

Handling views of an already existing XML layout:

- Suppose we have the following XML layout from one of our old projects:

```xml
<?xml version="1.0" encoding="utf-8"?>
<android.support.design.widget.CoordinatorLayout
    xmlns:android="http://schemas.android.com/apk/res/android"
    xmlns:app="http://schemas.android.com/apk/res-auto"
    xmlns:tools="http://schemas.android.com/tools"
    android:layout_width="match_parent"
    android:layout_height="match_parent"
tools:context="android.my_company.com.helloworldapp.HelloWorldActiv
ity">

    <android.support.design.widget.AppBarLayout
        android:layout_width="match_parent"
        android:layout_height="wrap_content"
        android:theme="@style/AppTheme.AppBarOverlay">

        <android.support.v7.widget.Toolbar
            android:id="@+id/toolbar"
            android:layout_width="match_parent"
            android:layout_height="?attr/actionBarSize"
            android:background="?attr/colorPrimary"
            app:popupTheme="@style/AppTheme.PopupOverlay" />

    </android.support.design.widget.AppBarLayout>

    <LinearLayout
        xmlns:android="http://schemas.android.com/apk/res/android"
        xmlns:app="http://schemas.android.com/apk/res-auto"
        xmlns:tools="http://schemas.android.com/tools"
        android:layout_width="match_parent"
        android:layout_height="match_parent"
        android:orientation="vertical"
        android:background="@color/white"
app:layout_behavior="@string/appbar_scrolling_view_behavior">

        <EditText
            android:id="@+id/name"
            android:layout_width="match_parent"
            android:layout_height="wrap_content"
            android:hint="What is your name?"/>

        <EditText
            android:id="@+id/message"
```

```
                android:layout_width="match_parent"
                android:layout_height="wrap_content"
                android:hint="Your message"/>

            <Button
                android:id="@+id/btn_send"
                android:layout_width="match_parent"
                android:layout_height="wrap_content"
                android:text="Send"/>

        </LinearLayout>

    </android.support.design.widget.CoordinatorLayout>
```

- We can use Anko to access the views from this XML layout and also get/set properties of these views. Check out this code:

```
    override fun onCreate(savedInstanceState: Bundle?) {
        super.onCreate(savedInstanceState)
        setContentView(R.layout.activity_hello_world)

        var toolbar = find<Toolbar>(R.id.toolbar)
        setSupportActionBar(toolbar)

        var name = find<EditText>(R.id.name)
        var msg = find<EditText>(R.id.message)
        var buttonSend = find<Button>(R.id.btn_send)

        buttonSend.onClick {
            toast("Hello, ${name.text} we have recorded your message!")
        }
    }
```

This is how our layout looks:

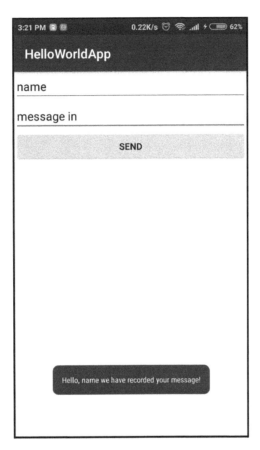

Logging using Anko

Logging is a great way to debug your application. You might have used
`android.util.Log`, which wasn't a very convenient way to log messages as it required
you to provide the `Log` tag with every message and also required you to define the tag,
which was usually the class name every time. Anko provides the anko-logger, which comes
with anko-commons. It is a very convenient way of logging messages as it doesn't require
you to necessarily override the log tag. In this recipe, we will learn how to do it.

Getting ready

I'll be using Android Studio to write the code. You need to add `anko-commons` to your `build.gradle` file. Anko logger comes within the `anko-commons` library:

```
dependencies {
    compile "org.jetbrains.anko:anko-commons:$anko_version"
}
```

How to do it...

Follow the given steps to learn how to use logging with the help of Anko library:

1. Logging in to Anko is very simple. You just need to implement AnkoLogger, as follows:

    ```
    class MainActivity : AppCompatActivity(),AnkoLogger {
    ```

2. Then, you can log messages, as follows:

    ```
    info("info message")
    ```

3. The following are the various levels of logging and their comparison with `android.Log.util`:

android.util.Log	AnkoLogger
v()	verbose()
d()	debug()
i()	info()
w()	warn()
e()	error()
wtf()	wtf()

4. The default tag name is a class name. If you want to override the `log` tag, you need to override the `loggertag` property:

    ```
    class MainActivity : AppCompatActivity(),AnkoLogger {
        override val loggerTag="CustomTag"
    ```

5. You can also use logger as a plain object. The following is an example from the documentation that uses logger as a plain object:

```
class SomeActivity : Activity() {
    private val log = AnkoLogger<SomeActivity>(this)
    private val logWithASpecificTag = AnkoLogger("my_tag")

    private fun someMethod() {
        log.warning("Big brother is watching you!")
    }
}
```

6. Each method has two versions: plain and lazy (inlined):

```
info("info message")
info{"info message"}
```

7. The lazy version is executed if `Log.isLoggable(tag, Log.INFO)` is true.

Handling dimensions with Anko

In XML, we use `dp` or `dip` as **density independent pixels** for layouts and views and `sp` as scale independent pixels for text. `dp` is a virtual pixel used to define layout sizes in a density-independent way; `sp` is like `dp`, but it's also scaled according to the user's font preference. In this recipe, we will understand how we can define dimensions of views and text in `dp` and `sp` in DSL layouts.

Getting ready

I'll be using Android Studio 3 to write code. You can get started by creating a new project in Kotlin with a blank activity in Android Studio 3+ as we won't be using any code from other recipes. You also need an intermediate understanding of Android development. Ensure that you have added Anko dependencies to your project by adding the following lines to your app-level `build.gradle` file and syncing the project:

```
compile "org.jetbrains.anko:anko:$anko_version"
```

Here, `$anko_version` is the latest version of Anko out there.

How to do it...

In the given steps, we will learn how to work with dimensions using Anko library:

1. Let's make a layout with a button of 120 `dip` width and `wrapContent` height and a text view with 24 `sp` text size. I suggest you to do this on your own, using this syntax:

```
dip(dipValue)
sp(spValue)
```

2. The following is one way you can create the layout using the `dip()` and `sp()` methods. `sp` is usually used for text, but to demonstrate, I have used it for a view height in the next example. Anko takes the value of the `textSize` property in `sp` by default, while you have to provide float:

```
verticalLayout {
    padding = dip(20)

    textView {
        text = "A big text view"
        textSize = 24f
    }

    button("Click me") {
        onClick {
            toast( "Hey! Here is a toast for you.")
        }
    }.lparams(dip(280), sp(80))

}
```

3. The following is how our layout looks:

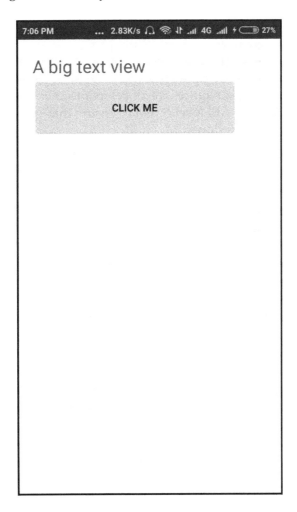

4. Anko also provides additional methods for our convenience, which are `px2dip(pixels)` and `px2sp(pixels)`, to convert pixels to dip and sp respectively. I remember coding them by hand before Anko existed, so they are handy a lot of times.

Version checking on Android

Android versions are shipped out very frequently. With every latest version of Android, you get new features and new improvements. Though Google tries very hard to provide backward compatibility, they aren't able to do so always. For example, there is no backward compatibility for Material design components; you need to be targeting API levels greater than 21 in order to use them. This requires the developer to check beforehand whether the component is supported on that API level or not to ensure that your app runs smoothly on all levels. We usually do that as follows:

```
if(Build.VERSION.SDK_INT>Build.VERSION_CODES.JELLY_BEAN){

}
```

Anko provides helper functions that help us achieve similar things but with easier syntax. In this recipe, we will see how to use it.

Getting ready

I'll be using Android Studio to write code. You also need to include the Anko library by adding these lines to your `build.gradle` file:

```
compile "org.jetbrains.anko:anko:$anko_version"
```

How to do it...

It's very easy to add version checks in Anko, and Anko provides two main functions for it:

- `doIfSdk` : This takes in the version code as a parameter, and also a function. If the API level of device is equal to the version code supplied, the function is executed. Here's an example of this function:

    ```
    doIfSdk(Build.VERSION_CODES.LOLLIPOP){
        // Do something specific to version 21
    }
    ```

- doFromSdk : This also takes in the version code as a parameter, along with the function, and executes that function if the device SDK level is greater than or equal to the supplied version code. The following is an example of the same:

```
doFromSdk(Build.VERSION_CODES.LOLLIPOP){
    // Execute this method on API >=21
}
```

How it works...

Let's see the implementation of the preceding two helper methods:

- For doIfSdk:

```
inline fun doIfSdk(version: Int, f: () -> Unit) {
    if (Build.VERSION.SDK_INT == version) f()
}
```

- For doFromSdk:

```
inline fun doFromSdk(version: Int, f: () -> Unit) {
    if (Build.VERSION.SDK_INT >= version) f()
}
```

As you can see, it's just old Android SDK code behind the hood. The doFromSdk and doIfSdk are just the syntactic sugars on top of it.

9
Anko Layouts

The following recipes will be covered in this chapter:

- Setting up Anko library for Anko layouts in Gradle

- Creating user-interface programmatically

- Working with the old code of XML layouts

- Using the provided AnkoComponent interface

- Setting theme for Android views in Anko

- Setting layout parameters for Anko views

- Adding listeners to Anko views

- Inserting XML layouts into DSL

- Converting XML files into DSL

- Showing Snackbar

- Showing Toasts

- Accessing views using synthetic properties

- Accessing views of view groups using extension functions

Introduction

Anko is a Kotlin library that makes Android development a lot faster and easier. It also makes the code clean and concise. Most of us are used to writing XML layouts for UI in Android, which is redundant and neither type safe nor null safe. It also eats CPU time and battery to parse XML on a device. Some who programmatically write layouts know how large the code becomes, and it is also very difficult to maintain.

With Anko, we can use DSL to define layouts. The advantages of using DSL is that they are easy to read and write and they have no runtime overhead. If you are familiar with Android development and XML layouts, this chapter will help you quickly get started with Anko layouts.

Setting up Anko library for Anko layouts in Gradle

The first and foremost thing to do to start using any library is adding its dependencies to our project so as to be able to use its methods and functionalities in our project. In this recipe, we will explore how to add dependencies of Anko layouts to our project using gradle.

Getting ready

I'll be using Android Studio 3 to write the code, as it is the latest right now. You can get started by creating a new project in Kotlin with a blank activity in Android Studio 3+ as we won't be using any code from other recipes. You also need an intermediate understanding of Android development.

How to do it...

In the following steps, we will add Anko to our project:

1. We can add all Anko features and components in one go by adding the following line to our `build.gradle` dependencies:

```
// Anko
compile "org.jetbrains.anko:anko:$anko_version"
```

Here, `$anko_version` is the latest version of Anko. You can replace it with the latest version of Anko at this time.

2. After that, sync your `build.gradle` file. Now, Anko dependencies have been added to your project. Let's check this by simply using Anko commons to create and show an alert dialog. Create a button in your activity by defining it in your XML layout and adding `onClickListener` on it, clicking on which should run the following code:

```
alert("This is my message from alert dialog", "An Alert!") {
    yesButton { toast("Thanks for clicking ok") }
    noButton {
        toast("Got it!") }
}.show()
```

3. If on clicking the button, an alert shows up that we successfully added Anko library in our project, this is how the alert dialog looks:

4. However, most of the time we just need to add a single feature of Anko to our project. For example, Anko layouts in this case. So let's try to add just Anko layouts library to our project. Remove the previous code from `build.gradle` and your `Activity` and let's start over.

5. Now add the following lines to your project's app-level `build.gradle` dependencies:

```
// Anko Layouts
compile "org.jetbrains.anko:anko-sdk25:$anko_version"
compile "org.jetbrains.anko:anko-appcompat-v7:$anko_version"
```

6. Sync your `build.gradle` and if there are no errors, you can now use Anko layouts in your project. At this point, we should also add dependencies of Anko coroutines, as we will obviously need listeners on our layouts. You can add those dependencies by adding the following lines to your `build.gradle` file:

```
// Coroutine listeners for Anko Layouts
compile "org.jetbrains.anko:anko-sdk25-coroutines:$anko_version"
compile "org.jetbrains.anko:anko-appcompat-v7-
coroutines:$anko_version"
```

7. Done! Now, let's check whether everything is working perfectly. To do that, let's add a basic DSL layout to our main activity. Check out the following code of the `onCreate()` method of our target activity:

```
override fun onCreate(savedInstanceState: Bundle?) {
    super.onCreate(savedInstanceState)
    verticalLayout {
        button("Hello World button!") {
            onClick { toast("Hello, World!") }
        }
    }
}
```

8. Now run the app on your phone; if the layout works correctly, that is, you have a button on your screen with text **HELLO WORLD BUTTON!**, then we have successfully added Anko layouts dependencies to our project. This is how our layout looks:

9. Also, on clicking on the button, we get a toast like this:

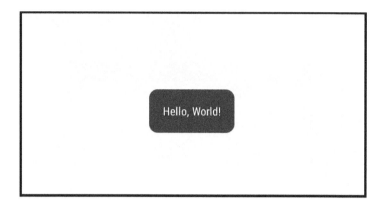

How it works...

By adding project dependencies in our `build.gradle` file, the amazing gradle takes care of what libraries and dependencies are needed by our project. The dependencies are located by gradle on our machine or in a remote repository, and any transitive dependencies are automatically included. Gradle makes adding project dependencies extremely easy and quick, and we can invest most of our time in creating our amazing software, rather than maintaining and resolving dependencies that becomes extremely difficult in large projects with lots of dependencies.

There's more...

If you do not use Gradle and do not want to use it in your project, you can directly add Anko's library JAR from the **jcenter repository** (`https://jcenter.bintray.com/org/jetbrains/anko/`) as library dependencies.

Creating user-interface programmatically

Writing UI in XML is not type safe or null safe, and it also eats CPU and battery. Writing UI programmatically (especially in Java) becomes bulky and unmanageable for large and complex UIs. That is when Anko layouts come to the rescue. We can easily create layouts in DSL using Anko layouts, and it also has no runtime overhead. In this recipe, we will see how to create layouts using DSL.

Getting ready

I'll be using Android Studio 3 to write code. You can get started by creating a new project in Kotlin with a blank activity in Android Studio 3+, as we won't be using any code from other recipes. You also need an intermediate understanding of Android development. Ensure that you have added Anko layouts dependencies to your project (follow the recipe *Setting up Anko library for Anko layouts in gradle*, in this chapter).

How to do it...

Let's start with a simple example where we use Anko to create a layout for our target activity (the activity in which you want to create a layout):

1. Here's the code for the `onCreate()` method that you need to put in your target activity:

```
override fun onCreate(savedInstanceState: Bundle?) {
    super.onCreate(savedInstanceState)
    verticalLayout {
        padding = dip(20)
        val name = editText {
            hint = "What is your name?"
        }
        val message = editText {
            hint = "Your message"
        }
        button("Send") {
            onClick { toast("Hello, ${name.text} we have recorded
your message!") }
        }
    }
}
```

2. Basically, in the preceding code, we want to create a basic "contact us" form. For this, we have created a vertical linear layout with a 20 dip padding and, inside the vertical linear layout, we have added two edit texts for name and message, respectively. On clicking on the button, we take the data and show the user a confirmation that the message has been recorded through toast. This is how part of the screen looks:

3. Anko layouts DSL is a great way to build UI in fewer lines of code. It is simple to read and write, and it's clean and concise. It has no runtime overhead as there is in XML layouts. Anko layouts support XML too, and you can use custom components and also use coroutines for listeners. You can also get a preview of the DSL layout in Android Studio when using `AnkoComponent` interface, which we will learn later in this chapter.

4. Let's try another example where we fit the preceding layout in a coordinator layout with a toolbar. To be able to use a coordinator layout, we need to add dependencies for Anko design support library. Add the following lines to your `build.gradle` and sync your project:

```
// Anko layouts design support
compile "org.jetbrains.anko:anko-design:$anko_version"
```

5. There are a lot of artifacts by Anko for various Android support libraries out there. The following is the list:

```
// Appcompat-v7 (only Anko Commons)
 compile "org.jetbrains.anko:anko-appcompat-v7-commons:$anko_version"
// Appcompat-v7 (Anko Layouts)
 compile "org.jetbrains.anko:anko-appcompat-v7:$anko_version"
 compile "org.jetbrains.anko:anko-coroutines:$anko_version"
// CardView-v7
 compile "org.jetbrains.anko:anko-cardview-v7:$anko_version"
// Design
 compile "org.jetbrains.anko:anko-design:$anko_version"
 compile "org.jetbrains.anko:anko-design-coroutines:$anko_version"
// GridLayout-v7
 compile "org.jetbrains.anko:anko-gridlayout-v7:$anko_version"
// Percent
 compile "org.jetbrains.anko:anko-percent:$anko_version"
// RecyclerView-v7
 compile "org.jetbrains.anko:anko-recyclerview-v7:$anko_version"
 compile "org.jetbrains.anko:anko-recyclerview-v7-coroutines:$anko_version"
// Support-v4 (only Anko Commons)
 compile "org.jetbrains.anko:anko-support-v4-commons:$anko_version"
// Support-v4 (Anko Layouts)
 compile "org.jetbrains.anko:anko-support-v4:$anko_version"
```

6. Now what we need is a coordinator layout that fits the whole width and height of the parent and, inside it, we need an app bar with a toolbar and below the app bar we need our vertical layout from earlier. I suggest that you try to code this one on your own before checking out my method, which is as follows:

```
coordinatorLayout {
    fitsSystemWindows = true
    lparams {
        width = matchParent
        height = matchParent
    }
```

```
appBarLayout {
    toolbar {
        setTitleTextColor(Color.WHITE)
        id = R.id.toolbar
        title = resources.getString(R.string.main_activity)
    }.lparams {
        width = matchParent
        height = wrapContent
    }
}.lparams { width = matchParent }
verticalLayout {
    verticalLayout {
        background =
context.getDrawable(R.color.colorLightGrey)
        gravity = Gravity.CENTER
        textView("logo"){
            textColor = context.getColor(R.color.colorAccent)
            textSize = 24f
        }.lparams(width = wrapContent, height = wrapContent) {
            horizontalMargin = dip(5)
            topMargin = dip(10)
        }
    }.lparams(width = matchParent, height = dip(200)) {
        horizontalMargin = dip(5)
        topMargin = dip(10)
    }
    padding = dip(20)
    val name = themedEditText(theme = R.style.newInput) {
        id = R.id.name
        hint = "What is your name?"
    }
    val message = editText {
        id = R.id.message
        hint = "Your message"
    }
    themedButton("Send", theme = R.style.newButton)
{
        id = R.id.btn_send
    }
}.lparams {
    width = matchParent
    height = matchParent
    behavior = AppBarLayout.ScrollingViewBehavior()
}
}
```

7. `lparams` used in the preceding function is the extension function used to add layout parameter to a view. This is how the layout looks in our app:

Creating layouts using DSL is a bit similar to XML itself, which is intentional, given the previous experience of developers with XML, and it also gives us the power to calculate things on the fly, while dynamically adding views.

How it works...

XML Parsing is done at compile time (except for a few things). It introduces CPU and battery overheads. For very complex layouts, it also introduces latency in the app and at times, severely affects user experience.

In Anko layouts, the DSL builds the layout at runtime and hence we can include anything. It also avoids runtime overhead, and we can avoid null pointer exception. Also, we do not need casting and can dodge the `findViewById` calls as well.

Working with the old code of XML layouts

The best thing about Anko layouts is the flexibility to be able to work with our XML layouts as well. Also, Anko makes things easier by providing us view properties. In this recipe, we will see how to use XML layouts and still be able to improve things using Anko Layouts.

Getting ready

I'll be using Android Studio 3 to write code. You can get started by creating a new project in Kotlin with a blank activity in Android Studio 3+ as we won't be using any code from other recipes. You also need an intermediate understanding of Android development. Ensure that you have added Anko layouts dependencies to your project (follow the recipe *Setting up Anko library for Anko layouts in gradle*, in this chapter).

How to do it...

In the following steps, we will learn how to work with XML layouts, along with Anko layouts:

1. Let's start by first having an old XML file to work on. Add the following code to an XML layout that you will add as the content view of your target activity:

```xml
<?xml version="1.0" encoding="utf-8"?>
<LinearLayout
xmlns:android="http://schemas.android.com/apk/res/android"
    xmlns:app="http://schemas.android.com/apk/res-auto"
    xmlns:tools="http://schemas.android.com/tools"
    android:layout_width="match_parent"
    android:layout_height="match_parent"
    app:layout_behavior="@string/appbar_scrolling_view_behavior"
tools:context="android.my_company.com.helloworldapp.MainActivity"
    tools:showIn="@layout/activity_main"
    android:orientation="vertical"
    android:padding="20dp">

    <EditText
        android:id="@+id/name"
        android:layout_width="match_parent"
        android:layout_height="wrap_content"
        android:hint="What is your name?"/>

    <EditText
```

```
        android:id="@+id/message"
        android:layout_width="match_parent"
        android:layout_height="wrap_content"
        android:hint="Your message"/>

    <Button
        android:id="@+id/btn_send"
        android:layout_width="match_parent"
        android:layout_height="wrap_content"
        android:text="Send"/>

</LinearLayout>
```

2. Traditionally, we used `findViewById()` and `onClickListener()` in our activity to manipulate properties of the elements of layout and handle events. However, with Anko layouts, this becomes as easy as the following:

```
override fun onCreate(savedInstanceState: Bundle?) {
    super.onCreate(savedInstanceState)
    setContentView(R.layout.activity_main)
    setSupportActionBar(toolbar)

    var name = find<EditText>(R.id.name)
    var msg = find<EditText>(R.id.message)
    var buttonSend = find<Button>(R.id.btn_send)

    buttonSend.onClick {
        toast("Hello, ${name.text} we have recorded your message!")
    }
}
```

3. The preceding is the `onCreate()` method of the target activity. Note that the `find()` method is a lot simpler than `findViewById()`.

4. We can get and set view properties and also attach listeners to view events. Another thing is that the Kotlin's Android extension functions also let us deal with views without using the `find` method. Check out the following code, whereby it becomes super easy to get and set view properties using synthetic extension properties:

```
override fun onCreate(savedInstanceState: Bundle?) {
    super.onCreate(savedInstanceState)
    setContentView(R.layout.activity_main)
    setSupportActionBar(toolbar)
    var nameText = name.text
    var msg = message.text
    btn_send.onClick {
```

```
                          toast("Hello, $nameText we have recorded your message!")
                }
        }
```

5. Here, `name`, `message`, and `btn_send` are the IDs of the views in XML layout, respectively.

How it works...

Anko provides us with these extension functions and properties that make it easier to access the views. Some of these functions and properties are prearranged into type-safe builders that are generated using Android JAR files.

There's more...

It is worth understanding how Kotlin's synthetic properties work. Kotlin generates some extra code that helps us use our views like properties, naming the variables similar to that of the ID of the view. Basically, it is running `findViewById()` the first time we try to access a view as property and storing it in cache so that all the consecutive calls to the same view invoke `findCachedViewById()`, thus making the access a lot faster.

Using the provided AnkoComponent interface

We can define an activity's layout DSL directly in the `onCreate()` method, but it is sometimes convenient to separate UI into another class. In this recipe, we will see how to use the `AnkoComponent` interface to do that.

Getting ready

I'll be using Android Studio 3 to write code. You can get started by creating a new project in Kotlin with a blank activity in Android Studio 3+ as we won't be using any code from other recipes. You also need an intermediate understanding of Android development. Ensure that you have added Anko layouts dependencies to your project (follow the recipe *Setting up Anko library for Anko layouts in gradle*, in this chapter).

How to do it...

In the given steps, we will learn how to work with the AnkoComponent interface:

1. Let's start by adding our UI in a different class that implements the `AnkoComponent` interface like this:

```
class MainActivityUI : AnkoComponent<MainActivity> {
    override fun createView(ui: AnkoContext<MainActivity>) =
with(ui) {
        verticalLayout {
            padding = dip(20)
            val name = editText {
                id = R.id.name
                hint = "What is your name?"
            }

            val message = editText {
                id = R.id.message
                hint = "Your message"
            }

            button("Send") {
                id = R.id.btn_send
            }
        }
    }
}
```

2. Note that the preceding class implements the `AnkoComponent` interface. We need to override the `createView()` method and return DSL layout from it. Now, let's see how to get this layout and set it to our activity. Check out the modified `onCreate()` method in our activity:

```
override fun onCreate(savedInstanceState: Bundle?) {
    super.onCreate(savedInstanceState)
    MainActivityUI().setContentView(this)
}
```

3. Now, let's try to access these views inside our activity, which we should be able to do if the layout has been properly set to our activity. We will access them just as we would have accessed a view from XML layout:

```
override fun onCreate(savedInstanceState: Bundle?) {
    super.onCreate(savedInstanceState)
    MainActivityUI().setContentView(this)
    btn_send.onClick { toast("Hello, ${name.text} we have recorded
```

```
your message!") }
}
```

4. The next image is how the screen will look:

5. Also, on entering the details and clicking on the button, we see the toast, as follows:

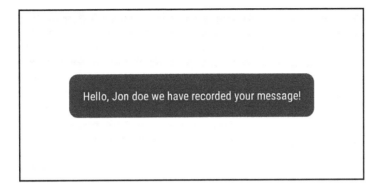

How it works...

The `verticalLayout` (which is a vertical linear layout) block is an extension function provided by Anko, which creates a new view instance and adds it to the parent. There are such extension functions for every view in the Android framework. For example, we used button and edit text as well in the preceding example. We can also use it as `button()`, which accepts a string parameter for text on the button or `button{}` if we want to set any properties on that view.

There's more...

If we use the `AnkoComponent` interface for creating our DSL in another class, we can also preview our layout DSL using the Anko support plugin.

For doing so, first add Anko support plugin from plugins in Android Studio settings. After that, put the cursor somewhere inside the `MainActivityUI` declaration, open the Anko Layout Preview tool window by clicking on **View|Tool Windows|Anko Layout Preview**, and press **Refresh**.

If the layout preview is not being rendered properly, rebuild the project. This is how the window looks:

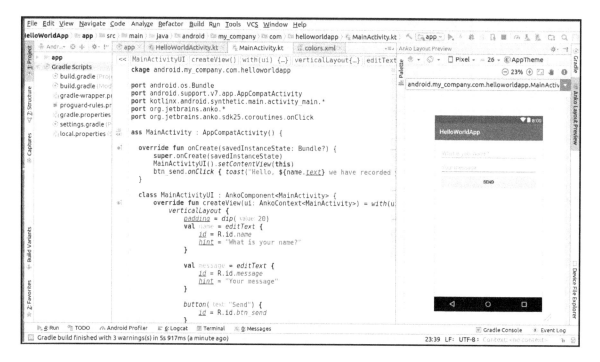

Setting theme for Android views in Anko

Our Android apps won't be so beautiful if we won't be able to style our views. Anko layouts give us the power to apply custom themes to our views. In this recipe, we will learn how to create themed views in Anko.

Getting ready

I'll be using Android Studio 3 to write code. You can get started by creating a new project in Kotlin with a blank activity in Android Studio 3+, as we won't be using any code from other recipes. You also need an intermediate understanding of Android development. Ensure that you have added Anko layouts dependencies to your project (follow the recipe *Setting up Anko library for Anko layouts in gradle*, in this chapter).

How to do it...

In the given steps, we will learn how to set the theme for Android views using Anko:

1. Let's start by first creating a style for a button. Custom styles are created in `styles.xml` inside the `res/values/` directory. Let's create a button style and name it `newButton`. Add the following code in `styles.xml`:

```
<style name="newButton" parent="android:Widget.Holo.Light.Button">
    <item
name="android:colorButtonNormal">@color/colorAccent</item>
    <item name="android:textColor">@color/white</item>
</style>
```

2. Now, let's use this style to create a themed button in our target activity. Let's keep our UI in another class using the `AnkoComponent` interface. The following is how we create a button with a custom theme in a DSL layout (focus on the bold parts of the code):

```
class MainActivityUI : AnkoComponent<MainActivity> {
    override fun createView(ui: AnkoContext<MainActivity>) =
with(ui) {
        verticalLayout {
            padding = dip(20)
            val name = editText {
                id = R.id.name
                hint = "What is your name?"
            }

            val message = editText {
                id = R.id.message
                hint = "Your message"
            }

            themedButton("Send", theme = R.style.newButton)
{

                id = R.id.btn_send
            }
        }
    }
}
```

3. Also, to set this layout of our activity, we add the
   `LayoutActivity().setContentView(this)` line in the `onCreate()` method,
   as follows (focus on the bold parts of the code):

```
override fun onCreate(savedInstanceState: Bundle?) {
        super.onCreate(savedInstanceState)
        MainActivityUI().setContentView(this)
        btn_send.onClick { toast("Hello, ${name.text} we have
recorded your message!") }
    }
```

4. The following is how the top part of our app screen looks, with themed button
   taking background color as the accent color as defined in our `colors.xml` file in
   the `res/values/` directory. The text color is white, just as we set in our custom
   style:

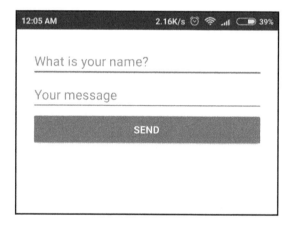

5. This is how we set themes on views, by appending themed keywords before the
   view name and making it camel case. We pass the theme as a parameter to the
   function.

Themed views are also Kotlin extension functions provided by Anko layouts.

Setting layout parameters for Anko views

Without layout parameters, there is not much we can do with our layouts. In this recipe, we
will see how to use layout parameters with views in our layout DSL.

Getting ready

I'll be using Android Studio 3 to write code. You can get started by creating a new project in Kotlin with a blank activity in Android Studio 3+, as we won't be using any code from other recipes. You also need an intermediate understanding of Android development. Ensure that you have added Anko layouts dependencies to your project (follow the recipe *Setting up Anko library for Anko layouts in gradle*, in this chapter).

How to do it...

In the following steps, we will learn how to set the layout parameters for Anko views:

1. Let's start with creating our view in an external class that inherits from the `AnkoComponent` interface. To add layout parameters to a view (which we add using the extension functions provided by Anko), we use the `lparams()` extension function, which we add at the end of our view block in DSL, and is something like this:

```
val message = editText {
    id = R.id.message
    hint = "Your message"
}.lparams() {
    // We specify our layout parameters here
}
```

2. Let's try a simple example with vertical layouts; check out the following code block (focus on the bold parts of the code):

```
verticalLayout {
    verticalLayout {
        background = context.getDrawable(R.color.colorLightGrey)
        gravity = Gravity.CENTER
        textView("logo"){
            textColor = context.getColor(R.color.colorAccent)
            textSize = 24f
        }.lparams(width = wrapContent, height = wrapContent) {
            horizontalMargin = dip(5)
            topMargin = dip(10)
        }

    }.lparams(width = matchParent, height = dip(200)) {
        horizontalMargin = dip(5)
        topMargin = dip(10)
    }
```

```
        padding = dip(20)
        val name = themedEditText(theme = R.style.newInput) {
            id = R.id.name
            hint = "What is your name?"
        }

        val message = editText {
            id = R.id.message
            hint = "Your message"
        }

        themedButton("Send", theme = R.style.newButton) {
            id = R.id.btn_send
        }
    }
}
```

3. The `themedEditText` and `themedButton` extension functions are provided by Anko to create an edit text and button with theme. If you do not wish to use themed views, simply call `editText()` and `button()` without passing the theme as a parameter.

4. Let's go through another example where we have a toolbar with the title of the page in it. Check out the next example that uses coordinator layout, app bar layout, and toolbar. The given code generates the layout as shown in the screenshot following the code:

```
coordinatorLayout {
    fitsSystemWindows = true
    lparams {
        width = matchParent
        height = matchParent
    }
    appBarLayout {
        toolbar {
            setTitleTextColor(Color.WHITE)
            id = R.id.toolbar
            title = resources.getString(R.string.main_activity)
        }.lparams {
            width = matchParent
            height = wrapContent
        }
    }.lparams { width = matchParent }
    verticalLayout {
        verticalLayout {
            background =
context.getDrawable(R.color.colorLightGrey)
            gravity = Gravity.CENTER
            textView("logo"){
                textColor = context.getColor(R.color.colorAccent)
                textSize = 24f
            }.lparams(width = wrapContent, height = wrapContent) {
                horizontalMargin = dip(5)
                topMargin = dip(10)
            }
        }.lparams(width = matchParent, height = dip(200)) {
            horizontalMargin = dip(5)
            topMargin = dip(10)
        }
        padding = dip(20)
        val name = themedEditText(theme = R.style.newInput) {
            id = R.id.name
            hint = "What is your name?"
        }
        val message = editText {
            id = R.id.message
            hint = "Your message"
        }
        themedButton("Send", theme = R.style.newButton)
```

```
            {
                    id = R.id.btn_send
              }
        }.lparams {
                width = matchParent
                height = matchParent
                behavior = AppBarLayout.ScrollingViewBehavior()
          }
      }
```

The following is how the layout looks in our app:

How it works...

`lparams` is also an Anko extension function that is added to views, and we can define layout parameters as properties. If you omit width and/or height while using `lparams()`, their values automatically default to `wrapContent`, just like in XML. The parameters passed are named arguments. Some of the properties are `horizontalMargin`, `verticalMargin`, and `margin`. For different layouts, we have different layout parameters, just as in XML. For example, for relative layout, we have `alignParentBottom()`, `alignParentTop()`, `alignParentStart()`,`leftOf(view IdOfReferenceView)`,`topOf(viewIdOfReferenceView)` and so on.

Check out the following example, which has the relative layout as root layout:

```
class MainActivityUI : AnkoComponent<MainActivity> {
    override fun createView(ui: AnkoContext<MainActivity>) =      with(ui) {
        relativeLayout {
            button("Ok") {
                id = R.id.ok
            }.lparams { leftOf() }

            button("Cancel").lparams { leftOf(R.id.ok) }
            lparams(matchParent, matchParent)
        }
    }
}
```

This is how the preceding layout looks:

Adding listeners to Anko views

We have event listeners on views in Android. Let's understand how Anko makes this easier by providing us with listener helpers.

Getting ready

I'll be using Android Studio 3 to write code. You can get started by creating a new project in Kotlin with a blank activity in Android Studio 3+, as we won't be using any code from other recipes. You also need an intermediate understanding of Android development. Ensure that you have added Anko layouts dependencies to your project (follow the recipe *Setting up Anko library for Anko layouts in gradle*, in this chapter).

How to do it...

In the following steps, we will learn how to add an event listener to Anko views:

1. Let's start with a simple example where we listen for click events on a button. Here's the code for attaching an `onClick` listener on a button with the `btn_send` ID:

```
btn_send.onClick { toast("Hello there we have recorded your
message!") }
```

2. The preceding code is the same as this:

```
var btn = find<EditText>(R.id.btn_send)
btn.setOnClickListener(object : OnClickListener {
    override fun onClick(v: View) {
      toast("Hello there we have recorded your message!")
    }
})
```

3. Now, let's create a layout having a button and a rating bar. We will attach an `onLongPress` listener on the button and an `onRatingBarChange` listener on the rating bar. Check out this code:

```
verticalLayout {
    padding = dip(20)
    val name = editText {
        id = R.id.name
        hint = "What is your name?"
    }

    val message = editText {
        id = R.id.message
        hint = "Your message"
    }
```

```
button("Send") {
    id = R.id.btn_send
    onLongClick {
        toast("Hello there we have recorded your message!")
    }
}

var rating = ratingBar {
    id = R.id.rating_bar
    onRatingBarChange { ratingBar, rating, fromUser ->
        toast(rating.toString())
    }
}.lparams(wrapContent, wrapContent)
}
```

Focus on the bold text in the preceding code. We can attach listeners by directly putting them inside the defined views. This is how our layout looks:

4. On long-pressing the button in the preceding screen with text SEND, we see a toast just as expected. Check out the following screen:

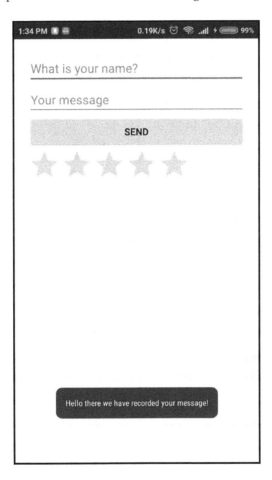

5. Similarly, if we choose a rating from the rating bar, we get a toast for the rating that we chose.

6. We can also keep the listeners separate from the layout, as in the following code. However, we do need the ID of the view we are setting the listener on for this to work:

```
override fun onCreate(savedInstanceState: Bundle?) {
    super.onCreate(savedInstanceState)
    MainActivityUI().setContentView(this)
    btn_send.onLongClick {
        toast("Hello there have recorded your message!")
    }
}
```

How it works...

Anko provides event listeners as extension functions to help ease the process of adding event listeners. We can also pass coroutines to these listener helpers and partially define listeners that have a lot of methods, that is, we can define each listener method separately and then they are merged by Anko if they are on the same view.

There's more...

Coroutines are used for writing asynchronous non-blocking code. You can also say that coroutines are threads managed by the user.

Inserting XML layouts into DSL

Sometimes a situation may arise where we might need to include an XML layout inside a DSL layout. Anko provides a solution for this. In this recipe, we will understand how to include XML layouts into DSL.

Getting ready

I'll be using Android Studio 3 to write code. You can get started by creating a new project in Kotlin with a blank activity in Android Studio 3+ as we won't be using any code from other recipes. You also need an intermediate understanding of Android development. Ensure that you have added Anko layouts dependencies to your project (follow the recipe *Setting up Anko library for Anko layouts in gradle*, in this chapter).

How to do it...

In the following steps, we will learn how to insert an XML layout into a DSL layout:

1. To include an XML layout in a DSL, we use the `include()` method. We can add view properties to the view created using the `include()` method by simply adding `{}` and defining our view properties inside it. We can also add layout parameters to the view, just as we do it in DSL views. Check out the syntax given here:

```
include<View>(R.layout.layoutName) {
    id = R.id.someId
    hint = "Some hint"
    text = "Some text"
}.lparams() {}
```

2. Let's create a layout in XML, which we will then include in our DSL layout. Let's create a button in a linear layout and save it in a file called `test.xml`. Check out the following code for the layout that we will save in `text.xml`:

```xml
<?xml version="1.0" encoding="utf-8"?>
<LinearLayout
xmlns:android="http://schemas.android.com/apk/res/android"
    android:orientation="vertical"
    android:layout_width="wrap_content"
    android:layout_height="wrap_content"
    android:padding="10dp">

    <Button
        android:id="@+id/btn_test"
        android:layout_width="wrap_content"
        android:layout_height="wrap_content"
        android:text="Send"
        android:background="@color/colorAccent"
        android:textColor="@color/white"/>
</LinearLayout>
```

3. The following is how our `test.xml` XML layout looks, a button with `10dp` space around it:

4. Now you need to try to include the layout you just created in a DSL layout on your own. You can add the DSL layout to the activity's `onCreate()` method or in an external class that implements the `AnkoComponent` interface. Check out the following code for DSL layout (focus on the bold text in the given code):

```
verticalLayout {
    padding = dip(20)
    val name = editText {
        id = R.id.name
        hint = "What is your name?"
    }

    val message = editText {
        id = R.id.message
        hint = "Your message"
    }

    button("Send") {
        id = R.id.btn_send
        onClick {
            toast("Hello there we have recorded your message!")
        }
    }

    include<View>(R.layout.test) {
        backgroundColor = Color.CYAN
    }.lparams(width = matchParent) { }
}
```

This is how our layout looks after we include `test.xml` in our DSL:

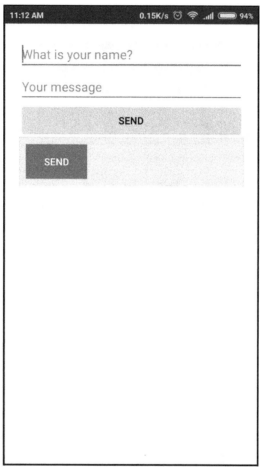

5. We can attach listeners and get/set properties of included views by accessing them using Kotlin's synthetic properties, using Anko's `find()` method, or using the `findViewById()`. Each of the preceding methods needs the view to have an ID. Check out the following code for attaching an on-click listener on the button in `test.xml` that has a `btn_test` ID:

```
override fun onCreate(savedInstanceState: Bundle?) {
        super.onCreate(savedInstanceState)
        MainActivityUI().setContentView(this)
        btn_test.onClick {
            toast("test click")
```

```
        }
    }
```

6. I have imported all views from `test.xml` by importing the synthetic properties of `test.xml`, as shown:

```
import kotlinx.android.synthetic.main.test.*
```

Converting XML files into DSL

If you are already an Anko fan and would love to port your old projects' XML to DSL without doing it manually, then this recipe will help you learn how to go about doing that.

Getting ready

I'll be using Android Studio 3 to write code. You can get started by creating a new project in Kotlin with a blank activity in Android Studio 3+, as we won't be using any code from other recipes. You also need an intermediate understanding of Android development. Ensure that you have added Anko layouts dependencies to your project (follow the recipe *Setting up Anko library for Anko layouts in gradle*, in this chapter).

How to do it...

Let's start by creating a blank activity and working on the XML layout to have something to convert to DSL. I have the following XML layout that I will convert to DSL next:

```
<?xml version="1.0" encoding="utf-8"?>
<android.support.design.widget.CoordinatorLayout
xmlns:android="http://schemas.android.com/apk/res/android"
    xmlns:app="http://schemas.android.com/apk/res-auto"
    xmlns:tools="http://schemas.android.com/tools"
    android:layout_width="match_parent"
    android:layout_height="match_parent"
    tools:context="android.my_company.com.helloworldapp.Main2Activity">

    <android.support.design.widget.AppBarLayout
        android:layout_width="match_parent"
        android:layout_height="wrap_content"
        android:theme="@style/AppTheme.AppBarOverlay">

        <android.support.v7.widget.Toolbar
```

```
            android:id="@+id/toolbar"
            android:layout_width="match_parent"
            android:layout_height="?attr/actionBarSize"
            android:background="?attr/colorPrimary"
            app:popupTheme="@style/AppTheme.PopupOverlay" />

    </android.support.design.widget.AppBarLayout>

    <LinearLayout
        android:layout_width="match_parent"
        android:layout_height="match_parent"
        android:orientation="vertical"
        android:gravity="center">

        <TextView
            android:id="@+id/text1"
            android:text="@string/hello_calendar"
            android:layout_width="wrap_content"
            android:layout_height="wrap_content"
            android:layout_margin="@dimen/dp10"
            style="@style/TextAppearance.AppCompat.Title"/>

        <CalendarView
            android:id="@+id/calendarView"
            android:layout_width="match_parent"
            android:layout_height="180dp"
            android:layout_margin="@dimen/dp10"/>

        <Button
            android:id="@+id/btn_done"
            android:background="@color/colorAccent"
            android:text="@string/done"
            android:layout_width="wrap_content"
            android:layout_height="wrap_content"
            android:layout_margin="@dimen/dp10"/>
    </LinearLayout>

    <android.support.design.widget.FloatingActionButton
        android:id="@+id/fab"
        android:layout_width="wrap_content"
        android:layout_height="wrap_content"
        android:layout_gravity="bottom|end"
        android:layout_margin="@dimen/fab_margin"
        app:srcCompat="@android:drawable/ic_dialog_email" />

</android.support.design.widget.CoordinatorLayout>
```

Showing Snackbar

Snackbars are a great way to show feedback and messages to the users. Snackbars show a message at the bottom of a mobile or lower-left on larger devices. They can also have an action button. They automatically disappear after the timeout or after user interaction or if the user swipes on the snackbar.

In this recipe, we will learn how to easily show a Snackbar using Anko layouts. Showing Snackbars in the traditional way is a bit long; Anko makes it simpler to quickly show snackbars. Let's see how.

Getting ready

I'll be using Android Studio 3 to write code. You can get started by creating a new project in Kotlin with a blank activity in Android Studio 3+, as we won't be using any code from other recipes. You also need an intermediate understanding of Android development. Ensure that you have added Anko layouts dependencies to your project (follow the recipe *Setting up Anko library for Anko layouts in gradle,* in this chapter).

How to do it...

In the below steps, we will learn how to show a snackbar using Anko library:

1. Let's create a few buttons, each for different snackbars. We will create a snackbar inside the `onClick` listener of each button. Here are the syntaxes of some snackbars. I suggest you try to code this on your own before moving to the solution:

```
snackbar(parentView, "feedback message")
snackbar(parentView, R.string.message_string)
longSnackbar(parentView, "longer message")
snackbar(parentView, "message for action snackbbar", "Action name")
{ doSomething() }
```

2. Check out one possible solution:

```
verticalLayout {
    id = R.id.rootView
    padding = dip(20)
    button("Simple Snackbar") {
        id = R.id.btn_snack1
        onClick {
            snackbar(rootView, "Hey! I'm a simple snackbar.")
        }
    }

    button("Simple Snackbar using resources") {
        id = R.id.btn_snack2
        onClick {
            snackbar(rootView, R.string.snack_message)
        }
    }

    button("Long Snackbar") {
        id = R.id.btn_snack3
        onClick {
            longSnackbar(rootView, R.string.snack_message)
        }
    }

    button("Action Snackbar") {
        id = R.id.btn_snack3
        onClick {
            longSnackbar(rootView, "Simple action snackbar rocks.",
"Action")
            {
                toast("Let us do some stuff!")
            }
        }
    }
}
```

This is how the layout looks:

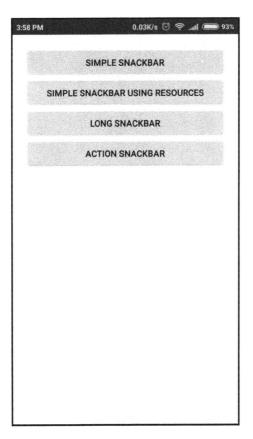

The following screenshot is how a snackbar without an action button looks:

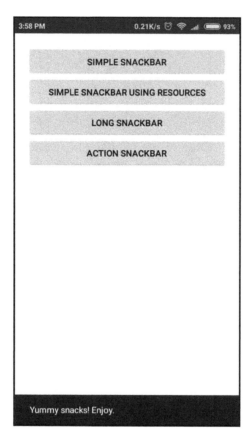

This is the one with an action button:

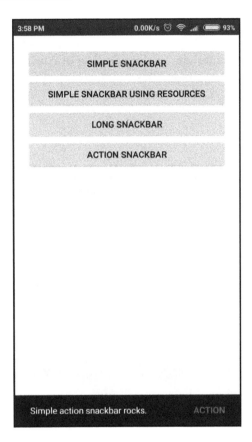

Showing Toasts

Toasts are used to show feedback or message as a popup in android. Toasts automatically disappear after their timeout. Showing toasts in Anko is extremely easy. Let's see how.

Getting started

I'll be using Android Studio 3 to write code. You can get started by creating a new project in Kotlin with a blank activity in Android Studio 3+, as we won't be using any code from other recipes. You also need an intermediate understanding of Android development. Ensure that you have added Anko layouts dependencies to your project (follow the recipe *Setting up Anko library for Anko layouts in gradle*, in this chapter).

How to do it...

Let's create a few buttons in our layout, clicking on them will show a toast:

- This is the syntax of toasts using Anko:

```
toast("a toast message")
toast(R.string.message_string)
longToast("a long duration toast message")
```

I suggest that you try showing toasts on the click of a button on your own, before moving on to the solution. Let's make a layout with three buttons that show toast on clicking using the preceding syntaxes.

The following is one way of creating a layout with three buttons where we have put our code to show toast inside onClick listener of the buttons. You can also put your layout in an external class that implements the AnkoComponent interface:

```
override fun onCreate(savedInstanceState: Bundle?) {
        super.onCreate(savedInstanceState)
        verticalLayout {
            id = R.id.rootView
            padding = dip(20)

            button("Show toast") {
                id = R.id.btn_snack1
                onClick {
                    toast( "Hey! Here is a toast for you.")
                }
            }

            button("Show toast using resource") {
                id = R.id.btn_snack2
                onClick {
                    toast(R.string.toast_message)
                }
```

```
        }

    button("Show long toast") {
        id = R.id.btn_snack3
        onClick {
            longToast(R.string.toast_message)
        }
    }
  }
}
```

The following is how our layout looks, and how the toasts appear on clicking on buttons:

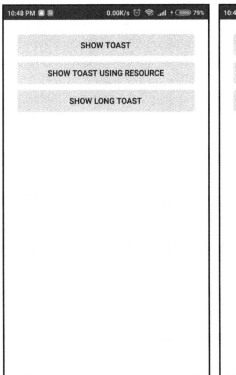

Accessing views using synthetic properties

So we know how Anko makes handling views and layouts easy, but Kotlin makes accessing views and getting/setting properties of views real fun. If you have tried using `findViewById()`, you already know how error-prone clunky code it is. There are a lot of libraries out there that provide a solution to this, but Kotlin provides a built-in plugin for this issue. Let's understand how to use that.

Getting ready

I'll be using Android Studio 3 to write code. You can get started by creating a new project in Kotlin with a blank activity in Android Studio 3+, as we won't be using any code from other recipes. You also need an intermediate understanding of Android development. Ensure that you have added Anko layouts dependencies to your project (follow the recipe *Setting up Anko library for Anko layouts in gradle*, in this chapter).

How to do it...

In the following steps, we will learn how to access views using synthetic properties:

1. Let's start with an XML layout and an activity that uses this XML layout. Start with creating a blank activity and create an XML layout that you wish to work with. I am working with the following layout file:

```xml
<?xml version="1.0" encoding="utf-8"?>
<android.support.design.widget.CoordinatorLayout
    xmlns:android="http://schemas.android.com/apk/res/android"
    xmlns:app="http://schemas.android.com/apk/res-auto"
    xmlns:tools="http://schemas.android.com/tools"
    android:layout_width="match_parent"
    android:layout_height="match_parent"
tools:context="android.my_company.com.helloworldapp.HelloWorldActiv
ity">

    <android.support.design.widget.AppBarLayout
        android:layout_width="match_parent"
        android:layout_height="wrap_content"
        android:theme="@style/AppTheme.AppBarOverlay">

        <android.support.v7.widget.Toolbar
            android:id="@+id/toolbar"
            android:layout_width="match_parent"
```

```
                android:layout_height="?attr/actionBarSize"
                android:background="?attr/colorPrimary"
                app:popupTheme="@style/AppTheme.PopupOverlay" />

        </android.support.design.widget.AppBarLayout>

        <LinearLayout
            xmlns:android="http://schemas.android.com/apk/res/android"
            xmlns:app="http://schemas.android.com/apk/res-auto"
            xmlns:tools="http://schemas.android.com/tools"
            android:layout_width="match_parent"
            android:layout_height="match_parent"
            android:orientation="vertical"
            android:background="@color/white"
    app:layout_behavior="@string/appbar_scrolling_view_behavior">


            <EditText
                android:id="@+id/name"
                android:layout_width="match_parent"
                android:layout_height="wrap_content"
                android:hint="What is your name?"/>

            <EditText
                android:id="@+id/message"
                android:layout_width="match_parent"
                android:layout_height="wrap_content"
                android:hint="Your message"/>

            <Button
                android:id="@+id/btn_send"
                android:layout_width="match_parent"
                android:layout_height="wrap_content"
                android:text="Send"/>

        </LinearLayout>

    </android.support.design.widget.CoordinatorLayout>
```

2. To use synthetic properties of a view, we need to import them inside the activity, as follows:

```
import kotlinx.android.synthetic.main.xml_layout_name.*
```

3. The following is how we can directly use view ID to provide reference to our view and get/set properties of the view:

```
override fun onCreate(savedInstanceState: Bundle?) {
    super.onCreate(savedInstanceState)
    setContentView(R.layout.activity_main2)
    setSupportActionBar(toolbar)

    btn_send.onClick {
        toast("Hey there ${name.text}. We have recorded your
message.")
    }
}
```

This is how our layout looks and works:

Accessing views of view groups using extension functions

We can use extension functions to add new behaviors to a class that we may not even have access to. We can also add extension functions to view groups. One such view group is recycler view. Let's see how we can access views of a recycler view using extension functions.

Getting ready

I'll be using Android Studio 3 to write code. You can get started by creating a new project in Kotlin with a blank activity in Android Studio 3+, as we won't be using any code from other recipes. You also need an intermediate understanding of Android development.

How to do it...

Kotlin has some operators that we can use on a class. We will be overloading one of these operators to get views of our view group:

1. We can access views of a view group by overloading the `get` operator like this:

```
operator fun ViewGroup.get(position: Int): View
{
    return getChildAt(position)
}
```

2. Now, in order to get a view from the view group, we can use either of the following methods:

```
val view = viewContainer.get(2)
// where 2 is the position for the view we want to access
```

3. Alternatively, use the following method because we used operator overloading, and `el.get(index)` matches with the array-like `el[index]` operation:

```
val view = viewContainer[2]
// where 2 is the position for the view we want to access
```

How it works...

Extension functions provide the ability to add new functionalities to a class without modifying the class or inheriting it or using any design pattern. Extension functions are resolved statically and bear no connection with the class they extend.

By operator overloading, Kotlin gives us the ability to provide implementations of a predefined set of operators. To overload an operator, we can use a member function or an extension function, which we used in the preceding case.

10
Databases and Dependency Injection

The following recipes will be covered in this chapter:

- Using SQLite database in Kotlin

- Creating database tables

- Injecting dependencies in Kotlin

- Reading data from database

- Converting database cursor into list of objects

- Using parseOpt for nullable objects

- Inserting data into database

- Creating singletons in Kotlin

- Using Dagger2 with Kotlin

- Using Butterknife with Kotlin

Introduction

When we develop an app, we should bear in mind those situations when the app won't be connected to the internet. The user might be in an elevator or there might not be any network coverage when they try to use the app. To provide a great user experience, we need to ensure that some parts our app work even when there isn't any network connection. To be able to do this, we need a persistent storage mechanism in our app. It can be achieved by either using shared preferences or using the database. Shared preferences can come in handy when we have small amounts of data such as the app's setting values. Databases are much more powerful for situations when we need to store structured data. In this chapter, we will learn how to use Android's built-in database SQLite and will also learn about dependency injection with Dagger2, which is considered among the best practices for developing a quality app.

Using SQLite database in Kotlin

SQLite is a relational database. Android comes with a built-in SQLite database. It is an open source SQL database and is widely used in Android apps. However, doing it in a raw manner is very time-consuming and eats up a lot of development and testing time. You have to work with cursors, iterate over them row by row, and wrap code in `try-finally`, and such. Of course, you can use libraries that provide ORM mapping, which makes it easier to deal with a SQLite database, but if the database is small, it is expensive and is generally overkill. Kotlin, with Anko, provides a really easy way to deal with SQLite database. So let's get to work and see how we can use SQLite database in Kotlin.

Getting ready

We'll be using Android Studio 3.0 for coding. First, we need to add anko-sqlite to our `build.gradle` file:

```
dependencies {
    compile "org.jetbrains.anko:anko-sqlite:$anko_version"
}
```

You can replace `$anko_version` with the latest version of the library.

How to do it...

Anko provides a wrapper around our built-in SQLite API, which helps eliminate a lot of boilerplate code and also adds safety mechanisms such as closing the database after the code execution is complete and more.

While implementing a SQLite database, the first step is to create the database helper class. In this case, we need the class to extend the ManagedSQLiteOpenHelper class instead of the SQLiteOpenHelper class, which we used to do. ManagedSQLiteOpenHelper is concurrency aware and closes the database at the end of query executions.
Check out the following code for a simple database helper that I am using for this example:

```
class DatabaseHelper(ctx: Context) : ManagedSQLiteOpenHelper(ctx,
"SupportDatabase", null, 1) {
    companion object {
        private var instance: DatabaseHelper? = null

        @Synchronized
        fun getInstance(context: Context): DatabaseHelper {
            if (instance == null) {
                instance = DatabaseHelper(context.applicationContext)
            }
            return instance!!
        }
    }

    override fun onCreate(db: SQLiteDatabase) {
        db.createTable("Requests", true,
                "id" to INTEGER + PRIMARY_KEY + UNIQUE,
                "name" to TEXT,
                "message" to TEXT)
    }

    override fun onUpgrade(db: SQLiteDatabase, oldVersion: Int, newVersion:
Int) {
        db.dropTable("Requests", true)
    }
}
```

So basically, in onCreate, we create tables and in onUpgrade, we upgrade tables.
I am creating a single table in my database, which is Requests. In the Requests table, we have the name, message, and id fields as primary keys.

We can provide access to the database by adding it as an extension property to the context. This allows access to the database by any class that requires context. The following code adds the database as an extension property to the context:

```
// Access property for Context
val Context.database: DatabaseHelper
    get() = DatabaseHelper.getInstance(getApplicationContext())
```

I added the preceding code to the same file as that of the database helper, outside the class.

Now, here's my code for the activity, where I have fields for a name and message and on pressing the **Enter** button, the details are stored in the database:

```
class MainActivity : AppCompatActivity() {

    override fun onCreate(savedInstanceState: Bundle?) {
        super.onCreate(savedInstanceState)
        MainActivityUI().setContentView(this)
        btn_send.onClick {
            database.use {
                insert("Requests",
                        "id" to 1,
                        "name" to name.text.toString(),
                        "message" to message.text.toString())
            }
        }
    }
}

class MainActivityUI : AnkoComponent<MainActivity> {
    override fun createView(ui: AnkoContext<MainActivity>) = with(ui) {
        verticalLayout {
            gravity = Gravity.CENTER
            padding = dip(20)

            textView {
                gravity = Gravity.CENTER
                text = "Enter your request"
                textColor = Color.BLACK
                textSize = 24f
            }.lparams(width = matchParent) {
                margin = dip(20)
            }

            val name = editText {
                id = R.id.name
                hint = "What is your name?"
            }
```

```
editText {
    id = R.id.message
    hint = "What is your message?"
    lines = 3
}

button("Enter") {
    id = R.id.btn_send
}
            }
        }
    }
}
```

Note the code in bold. So basically, we can perform operations on the database inside the use block. The database will be opened at the beginning of the use block and closed after its execution.

The following screenshot is how our layout looks:

Now try to put something in the database. Here's a screenshot of my database, and the insert operation worked:

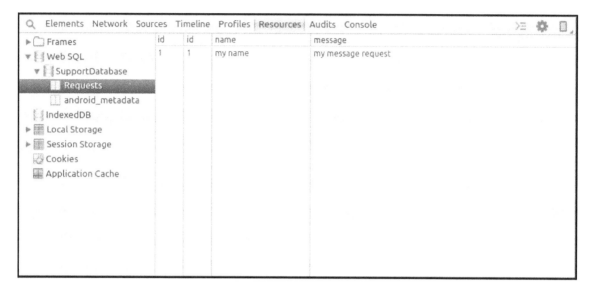

I am using Stetho (`https://github.com/facebook/stetho`) to view the database in Chrome dev tools.

For the layout of the activity, I have used Anko DSL layouts. You can refer to `Chapter 9`, *Anko Layouts*, of this book to learn more about them.

Creating database tables

Now that you have learned how to add anko-sqlite dependencies to your project and how to use SQLite database in the first recipe, the next step is learning how to create database tables.

Getting ready

We'll be using Android Studio 3 for coding. Ensure that you have added anko-sqlite to your `build.gradle` file and gone through the first recipe on how to use a SQLite database.

How to do it...

We will be creating two tables: `Requests` and `customers`:

1. For the `Requests` table, we have the `name` and `message` fields, and we can directly create them in the `onCreate` method of our database helper, as shown:

```
db.createTable("Requests", true,
    "id" to INTEGER + PRIMARY_KEY + UNIQUE,
    "name" to TEXT,
    "message" to TEXT)
```

2. For the `customers` table, we will be using a better coding practice by making a data class and using it to define the columns of the `customers` table. Given here is the code for our `Customer` data class:

```
data class Customer(val id: Int, val name: String, val phone_num:
String) {
    companion object {
        val COLUMN_ID = "id"
        val TABLE_NAME = "customers"
        val COLUMN_NAME = "name"
        val COLUMN_PHONE_NUM = "phone_num"
    }
}
```

3. Now, we will use this data class to create our table like this:

```
db.createTable(Customer.TABLE_NAME,
        true,
        Customer.COLUMN_ID to INTEGER + PRIMARY_KEY,
        Customer.COLUMN_NAME to TEXT,
        Customer.COLUMN_PHONE_NUM to TEXT)
```

4. The following is how our database helper finally looks after filling in the code for drop tables:

```
class DatabaseHelper(ctx: Context) : ManagedSQLiteOpenHelper(ctx,
"SupportDatabase", null, 1) {
    companion object {
        private var instance: DatabaseHelper? = null

        @Synchronized
        fun getInstance(context: Context): DatabaseHelper {
            if (instance == null) {
                instance =
DatabaseHelper(context.applicationContext)
            }
            return instance!!
        }
    }

    override fun onCreate(db: SQLiteDatabase) {
        db.createTable("Requests", true,
                "id" to INTEGER + PRIMARY_KEY + UNIQUE,
                "name" to TEXT,
                "message" to TEXT)

        db.createTable(Customer.TABLE_NAME,
                true,
                Customer.COLUMN_ID to INTEGER + PRIMARY_KEY,
                Customer.COLUMN_NAME to TEXT,
                Customer.COLUMN_PHONE_NUM to TEXT)
    }

    override fun onUpgrade(db: SQLiteDatabase, oldVersion: Int,
newVersion: Int) {
        db.dropTable("Requests", true)
        db.dropTable(Customer.TABLE_NAME, true)
    }
}

// Access property for Context
val Context.database: DatabaseHelper
    get() = DatabaseHelper.getInstance(getApplicationContext())
```

5. Now, let's do a fresh install of our app and see whether the two tables have been formed in our database. The following is how our database screenshot looks (using Stetho), and our tables have successfully been created:

Injecting dependencies in Kotlin

In Android development, Dagger 2 is the most popular dependency injection framework. You define the dependency objects, and with the help of a Dagger component, you inject it where you want. In this recipe, we will see how to inject the dependencies. We won't be going into the details of how to work with Dagger 2 in detail; for that, you can refer to the *Using Dagger2 with Kotlin* recipe in this chapter.

Getting ready

We will be using Android Studio 3.0 for this recipe. Ensure that you have its latest version.

How to do it...

When you've defined all the dependency objects you need in the module class, you can get the component. Let's take look at the mentioned steps:

1. To inject the object, you just need to add the @Inject annotation before the variable, then the object will be injected there. Let's look at the following example:

   ```
   @Inject
   lateinit var mPresenter:AddActivityMvpPresenter
   ```

   We have also used the lateinit modifier to void null checks before using the variable.

2. Another way to do it is by constructor injection. To understand it, let's take a look at the given code:

   ```
   @Module
   class AddActivityModule {
     @Provides @ControllerScope
     fun providesAddActivityPresenter(addActivityPresenter:
   AddActivityPresenter):AddActivityMvpPresenter =addActivityPresenter
   }
   ```

3. As you can see, we have sent AddActivityPresenter in the providesAddActivityPresenter, but the module doesn't provide it. This usually won't work unless you provide AddActivityPresnter as follows:

   ```
   class AddActivityPresenter @Inject constructor(var
   mDataManager:DataManager):AddActivityMvpPresenter
   ```

How it works...

When you use the @Inject annotation in the constructor, it means that the class needs the DataManager object before it can be created. Dagger2 will look into the dependency tree and provide you the dependency if it can.

Reading data from database

Now that we have seen how to create a database and how to create tables, let's learn how to read from a database.

Getting ready

I'll be using Android Studio 3 to write code. You can get started by adding anko-sqlite dependencies to your project and creating a SQLite database with the Requests table in it by going through and implementing the *Using SQLite database in Kotlin* recipe in this chapter. By using the form we created in this recipe, add some data to your Requests table.

How to do it...

Let's take look at the given steps to understand how to read data from the database:

1. Now, let's add a button to our existing layout from the first recipe; on clicking, it should retrieve all data from our Requests table. Check out the updated code, which is as follows, where I have added a button with a click listener on it:

```
class MainActivity : AppCompatActivity() {

    override fun onCreate(savedInstanceState: Bundle?) {
        super.onCreate(savedInstanceState)
        MainActivityUI().setContentView(this)
        val btn_send = find<Button>(R.id.btn_send)
        btn_send.onClick {
            database.use {
                insert("Requests",
                        "name" to name.text.toString(),
                        "message" to message.text.toString())
            }
            toast("success")
            name.text.clear()
```

```
                    message.text.clear()
        }
        val btn_read = find<Button>(R.id.btn_read)
        btn_read.onClick {
            var reqs = database.use {
    select("Requests").parseList(classParser<Request>())
            }
            for(x in reqs) {
                logd(x.name + ": " + x.message)
            }
        }
    }

    private fun logd(s: String) {
        Log.d("request", s)
    }

    class MainActivityUI : AnkoComponent<MainActivity> {
        override fun createView(ui: AnkoContext<MainActivity>) =
with(ui) {
            verticalLayout {
                padding = dip(20)

                textView {
                    gravity = Gravity.CENTER
                    text = "Enter your request"
                    textColor = Color.BLACK
                    textSize = 24f
                }.lparams(width = matchParent) {
                    margin = dip(20)
                }

                val name = editText {
                    id = R.id.name
                    hint = "What is your name?"
                }

                editText {
                    id = R.id.message
                    hint = "What is your message?"
                    lines = 3
                }

                button("Enter") {
                    id = R.id.btn_send
                }

                button("Show me requests") {
```

```
                                id = R.id.btn_read
                          }
                     }
                }
           }

        class Request(val id: Int, val name: String, val message:
   String)


   }
```

2. I am using Anko DSL to create the layout for my activity. As we discussed in previous recipes, we do all database operations inside the `database.use{...}` block. To read data from the database, we use the `select` function. The syntax is this:

```
db.select(tableName, vararg columns) // where db is an instance of
the SQLiteDatabase
```

3. Inside `database.use {...}`, `this` is the database instance, so we can directly use methods such as `select` and `insert`. The following is the data and the output in the database table:

This is the data:

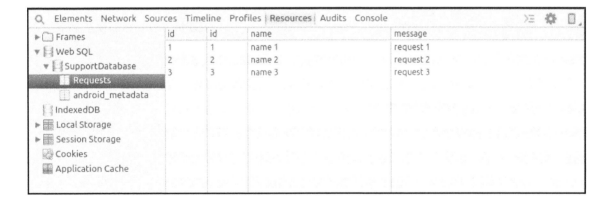

Here's the output:

```
11-18 18:21:34.709 12523-12523/android.my_company.com.helloworldapp
D/request: name 1: request 1
11-18 18:21:34.709 12523-12523/android.my_company.com.helloworldapp
D/request: name 2: request 2
11-18 18:21:34.709 12523-12523/android.my_company.com.helloworldapp
D/request: name 3 : request 3
```

4. There's a lot more we can do with the query builder; listed here are the methods provided by Anko:

   - `column(String)`: used to add a column to our `select` query
   - `distinct(Boolean)`: used to add distinct to the query
   - `whereArgs(String)`: used to specify the raw `where` string
   - `whereArgs(String, args)`: used to specify the `where` query and corresponding arguments
   - `whereSimple(String, args)` : used to specify a `where` query with the `?` marks and corresponding arguments for `?`
   - `orderBy(String, [ASC/DESC])` : used to specify a column for order by
   - `groupBy(String)`: used to specify a column for group by
   - `limit(count: Int)`: used to limit the number of rows returned by the query
   - `limit(offset: Int, count: Int)`: used to limit the number of rows returned by the query after an `offset`
   - `having(String)`: used to specify the raw `having` expression
   - `having(String, args)`: used to specify the raw `having` expression with arguments

5. Let's try another example. In this example, we will select data from a database using the `where` clause:

```
select("Requests")
    .whereArgs("(id > {userId})",
        "userId" to 1)
```

Here's the output of the preceding query:

```
11-18 21:11:04.328 18149-18149/android.my_company.com.helloworldapp
D/request: name 2: request 2
11-18 21:11:04.329 18149-18149/android.my_company.com.helloworldapp
D/request: name 3 : request 3
```

6. After getting the query results, we also need to parse the result. We get a cursor as a result from the query and using methods provided by Anko, we can easily parse them into regular classes. In the preceding example, we made a class named `Request`:

```
class Request(val id: Int, val name: String, val message: String)
```

7. The class has all the fields that we may get as columns from our query result cursor. The following are the methods that we can use for parsing results:
   - `parseSingle(rowParser):  T` : Parses exactly and only one row; if there's more than one row in the cursor, then it throws an exception
   - `parseOpt(rowParser):  T?` : Parses zero or one row, but if there's more than one row in the cursor, then it throws an exception
   - `parseList(rowParser):  List<T>` : Parses zero or more rows

We used `parseList` in the preceding example. You can pass row parsers or map parsers, and you can also use `classParser` of the type of your custom class, which passes a row parser, like this:

```
val rowParser = classParser<Person>()
```

Converting database cursor into list of objects

In the previous recipe, we learned how to query data from a database table. We receive a cursor as result of the query. In this recipe, we will learn how to use `parseList` to convert the cursor into a list of objects.

Getting ready

I'll be using Android Studio 3 to write code. You can get started by adding anko-sqlite dependencies to your project and creating a database helper like we did in the *Using SQLite database in Kotlin* recipe.

How to do it...

Follow these steps to convert the cursor into a list of objects:

1. Let's start by creating a `Customer` class as a model for our `customers` table:

```
data class Customer(val id: Int, val name: String, val phone_num:
String) {
    companion object {
        val COLUMN_ID = "id"
        val TABLE_NAME = "customers"
        val COLUMN_NAME = "name"
        val COLUMN_PHONE_NUM = "phone_num"
    }
}
```

2. Now we will write code to create the `customers` table inside the database helper class. Check out the following code:

```
class DatabaseHelper(ctx: Context) : ManagedSQLiteOpenHelper(ctx,
"SupportDatabase", null, 1) {
    companion object {
        private var instance: DatabaseHelper? = null

        @Synchronized
        fun getInstance(context: Context): DatabaseHelper {
            if (instance == null) {
                instance =
DatabaseHelper(context.applicationContext)
            }
            return instance!!
        }
    }

    override fun onCreate(db: SQLiteDatabase) {
        db.createTable(Customer.TABLE_NAME,
                true,
                Customer.COLUMN_ID to INTEGER + PRIMARY_KEY,
                Customer.COLUMN_NAME to TEXT,
```

```
                    Customer.COLUMN_PHONE_NUM to TEXT)
        }

        override fun onUpgrade(db: SQLiteDatabase, oldVersion: Int,
    newVersion: Int) {
            db.dropTable(Customer.TABLE_NAME, true)
        }
    }

    // Access property for Context
    val Context.database: DatabaseHelper
        get() = DatabaseHelper.getInstance(getApplicationContext())
```

3. Now, we will create a form to enter customers and a button that shows all customers in the database table using the `select` function. We will use the `parseList` method to get rows in the resulting cursor as a `List`. We need to pass in a row parser or map parser inside the `parseList` method. The easiest way of doing this is using `classParser` provided by Anko and using our `Customer` class constructor to get a row parser, like this:

```
var customers = database.use {
    select(Customer.TABLE_NAME)
    .parseList(classParser<Customer>())
}
```

I suggest that you try this exercise on your own before moving to the solution.

The following is my version of the activity that also contains DSL layout:

```
class MainActivity : AppCompatActivity() {

    override fun onCreate(savedInstanceState: Bundle?) {
        super.onCreate(savedInstanceState)
        MainActivityUI().setContentView(this)
        val name = find<EditText>(R.id.name)
        val phone = find<EditText>(R.id.phone)
        btn_send.onClick {
            database.use {
                insert(Customer.TABLE_NAME,
                        Customer.COLUMN_NAME to name.text.toString(),
                        Customer.COLUMN_PHONE_NUM to phone.text.toString())
            }
            toast("success")
            name.text.clear()
            phone.text.clear()
        }
```

```
            val btn_read = find<Button>(R.id.btn_read)
            btn_read.onClick {
                var customers = database.use {
                    select(Customer.TABLE_NAME)
                            .parseList(classParser<Customer>())
                }
                // customers is the list of objects which we can now iterate on
    to get individual values as objects of Customer class
                for(c in customers) {
                    debug(c.name + " (" + c.phone_num + ")")
                }
            }
        }

        private fun debug(s: String) {
            Log.d("customer", s)
        }

        class MainActivityUI : AnkoComponent<MainActivity> {
            override fun createView(ui: AnkoContext<MainActivity>) = with(ui) {
                verticalLayout {
                    padding = dip(20)

                    textView {
                        gravity = Gravity.CENTER
                        text = "Enter the customer"
                        textColor = Color.BLACK
                        textSize = 24f
                    }.lparams(width = matchParent) {
                        margin = dip(20)
                    }

                    val name = editText {
                        id = R.id.name
                        hint = "Name"
                    }

                    editText {
                        id = R.id.phone
                        hint = "Phone no."
                    }

                    button("Enter") {
                        id = R.id.btn_send
                    }

                    button("Show me customers") {
                        id = R.id.btn_read
```

```
            }

            button("Delete all customers") {
                id = R.id.btn_delete
            }
        }
      }
    }
  }
```

The result of the query, that is, `customers`, is the list of objects that we can now iterate on to get individual rows as objects of the `Customer` class.

Using parseOpt for nullable object

We use `parseList` when we get multiple rows in our cursor, but when we get only one row, we use `parseSingle` or `parseOpt`. However, what is the difference between `parseSingle` and `parseOpt`? In this recipe, we will understand the difference between both and when to use which one.

Getting ready

I'll be using Android Studio 3 to write code. You can get started by adding anko-sqlite dependencies to your project and creating a database helper, like we did in the *Using SQLite database in Kotlin* recipe. You will need to read and implement the previous recipe to be able to follow this recipe.

How to do it...

If you have read and implemented the previous recipe, you must already have a `customers` table in your database. Follow the mentioned steps to understand the difference between `parseSingle` and `parseOpt`:

1. In the previous recipe, we used `parseList` to get a list of rows as objects. If we need to get only a single row as an object, then we need to use `parseSingle`. The following is the syntax of `parseSingle`:

    ```
    parseSingle(rowParser): T
    ```

2. Now we use it in the following way in our previous code:

```
btn_read.onClick {
    var c = database.use {
        select(Customer.TABLE_NAME)
            .whereArgs("(id = {userId})",
            "userId" to 1)
            .parseSingle(classParser<Customer>())
    }
    debug(c.name + " (" + c.phone_num + ")")
}
```

3. We are using `parseSingle` because we will only get one row in the cursor, but if we get zero rows from the cursor, that is, we get an empty cursor, then we get an exception:

```
android.database.sqlite.SQLiteException: parseSingle accepts only
cursors with a single entry
```

However, what if there is a possibility of getting an empty cursor when we are expecting a cursor with a single row? It will always throw an exception, that is, when we use `parseOpt`; `parseOpt` accepts zero or one rows of cursors. Also, if `parseOpt` gets a null object, it handles the scenario accordingly by giving the value of `null` for each column. Basically, `parseOpt` is used for cursors that can be empty and objects that can be `null`.

The syntax for `parseOpt` is as follows:

`parseOpt(rowParser): T?` // The ? denotes that the object returned is nullable.

Here's how we will use it in our code:

```
btn_read.onClick {
    var c = database.use {
        select(Customer.TABLE_NAME)
            .whereArgs("(id = {userId})",
            "userId" to 1)
            .parseOpt(classParser<Customer>())
    }
    debug(c?.name + " (" + c?.phone_num + ")")
}
```

Now even if the cursor returned is empty, we do not get an exception, and `null` values are printed as output.

This is the output in case of an empty table:

```
11-18 21:11:04.329 18149-18149/android.my_company.com.helloworldapp
D/customer: null (null)
```

Inserting data into database

Inserting data into the database using Anko SQLite is a piece of cake. In this recipe, we will learn how to do that.

Getting ready

I'll be using Android Studio 3 to write code. You can get started by adding anko-sqlite dependencies to your project by adding the following lines to your `build.gradle` file:

```
dependencies {
    compile "org.jetbrains.anko:anko-sqlite:$anko_version"
}
```

You can replace `$anko_version` with the latest version of the library.

How to do it...

Let's insert data into our database by following the mentioned steps:

1. Let's start with our database helper, in which we will be creating a `Requests` table with the `name`, `message`, and `id` fields, as follows:

```
class DatabaseHelper(ctx: Context) : ManagedSQLiteOpenHelper(ctx,
"SupportDatabase", null, 1) {
    companion object {
        private var instance: DatabaseHelper? = null

        @Synchronized
        fun getInstance(context: Context): DatabaseHelper {
            if (instance == null) {
                instance =
DatabaseHelper(context.applicationContext)
            }
            return instance!!
```

```
            }
        }

    override fun onCreate(db: SQLiteDatabase) {
        db.createTable("Requests", true,
                "id" to INTEGER + PRIMARY_KEY + UNIQUE +
AUTOINCREMENT,
                "name" to TEXT,
                "message" to TEXT)
    }

    override fun onUpgrade(db: SQLiteDatabase, oldVersion: Int,
newVersion: Int) {
        db.dropTable("Requests", true)
    }
}

// Access property for Context
val Context.database: DatabaseHelper
    get() = DatabaseHelper.getInstance(getApplicationContext())
```

2. Now, let's create an activity with a form that takes the name and message and stores it in the database. I am using Anko DSL layouts for the layout of the activity:

```
class MainActivity : AppCompatActivity() {

    override fun onCreate(savedInstanceState: Bundle?) {
        super.onCreate(savedInstanceState)
        MainActivityUI().setContentView(this)
        btn_send.onClick {
            database.use {
                insert("Requests",
                        "name" to name.text.toString(),
                        "message" to message.text.toString())
            }
            toast("success")
            name.text.clear()
            message.text.clear()
        }
    }

    class MainActivityUI : AnkoComponent<MainActivity> {
        override fun createView(ui: AnkoContext<MainActivity>) =
with(ui) {
            verticalLayout {
                padding = dip(20)
```

```
textView {
    gravity = Gravity.CENTER
    text = "Enter your request"
    textColor = Color.BLACK
    textSize = 24f
}.lparams(width = matchParent) {
    margin = dip(20)
}

val name = editText {
    id = R.id.name
    hint = "What is your name?"
}

editText {
    id = R.id.message
    hint = "What is your message?"
    lines = 3
}

button("Enter") {
    id = R.id.btn_send
}
            }
        }
    }
}
```

3. Note the code in bold in the preceding code snippet. We will do all operations inside the `database.use {...}` block, because it is concurrency safe and also closes the database after execution of the block. If you have gone through the *Creating database tables* recipe, you will note that table creation and insertion is quite similar. The syntax is this:

```
db.insert(TABLE_NAME,
    COLUMN_NAME_1 to VALUE_1,
    COLUMN_NAME_2 to VALUE_2,
    COLUMN_NAME_3 to VALUE_3
)
```

This is our layout:

On entering the data, we can check whether our name and message are being stored in our database or not. I am using Stetho to view the database on my device.

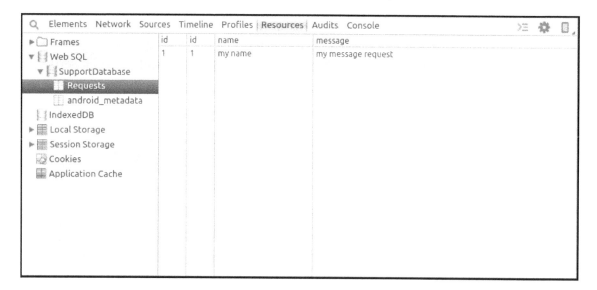

Creating singletons in Kotlin

A singleton class is a class that can have only one instance/object of that class at a time. The concept is to restrict instantiation of objects to a certain number. In this recipe, we will explore singletons in Kotlin.

Getting ready

I'll be using Android Studio 3 to write code.

How to do it...

Follow these steps to create singletons in Kotlin:

1. Kotlin does not have static members or variables, so for declaring static members of a class, we use `companion object`. Check out this example:

```
class SomeClass {

    companion object {
        var intro = "I am some class. Pleased to meet you!"
        fun infoIntro(): String {
            return "I am some class. Pleased to meet you!"
        }
    }
}
```

2. Accessing the members and methods of `companion` object of the preceding class is the same as we would do for any static members or methods:

```
var x = SomeClass.intro
toast(SomeClass.infoIntro())
```

3. Now what if we want a singleton class, that is, the class with only one object/instance at a time? Brace yourselves, this one is fun. Here's a way to create a singleton class in just a few lines:

```
object SomeClass {

    var intro = "I am some class. Pleased to meet you!"
    fun infoIntro(): String {
        return "I am some class. Pleased to meet you!"
```

```
        }
    }
```

Also, we use it just like the static members in the preceding examples:

```
var x = SomeClass.intro
toast(SomeClass.infoIntro())
```

How it works...

In Kotlin, decompiling bytecode is always a great way to know what is happening behind the scenes. If we decompile the bytecode of the object we created, we get the following code, which shows that behind the scenes, the object is just a class with a single instance at a time:

```
public final class SomeClass {
    @NotNull
    private static String intro;
    public static final SomeClass INSTANCE;

    @NotNull
    public final String getIntro() {
        return intro;
    }

    public final void setIntro(@NotNull String var1) {
        Intrinsics.checkParameterIsNotNull(var1, "<set-?>");
        intro = var1;
    }

    @NotNull
    public final String infoIntro() {
        return "I am some class. Pleased to meet you!";
    }

    private SomeClass() {
        INSTANCE = (SomeClass)this;
        intro = "I am some class. Pleased to meet you!";
    }

    static {
        new SomeClass();
    }
}
```

Using Dagger 2 with Kotlin

Dagger 2 is the best dependency injection framework in the Android community and is also open source. It is backed by Google and is widely used. Dependency injection is considered best practice and makes your code base scalable. In this recipe, we will learn how to use Dagger 2 for dependency injection in Kotlin.

Getting ready

We'll be using Android Studio 3.0 for coding purposes. First, we need to include Dagger 2 in the project, by adding the following lines to the `build.gradle` file:

```
compile "com.google.dagger:dagger:$daggerVersion"
kapt "com.google.dagger:dagger-compiler:$daggerVersion"
```

You need to replace `$daggerVersion` with the latest version of Dagger2.

How to do it...

Before we move ahead, we need to understand how Dagger2 works. Dagger2 uses annotation to generate codes and uses it to access fields; therefore, it can't use private fields.

The following annotations are used in Dagger2:

- `@Module` and `@Provides`: Define classes and methods that provide dependencies
- `@Inject`: Requests dependencies, and can be used on a constructor, a field, or a method
- `@Component`: Enables selected modules and is used for performing dependency injection

The classes annotated with `@Module` are responsible for providing objects that can be injected. The methods that provide those objects need to be annotated with `@Provides`. If the method requires another object to create a dependency object, they are provided in the method parameters. Dagger2 creates a dependency tree and checks whether the parameters can be supplied or not. Let's take a look at the implementation of a module:

1. We will look at an example of a network module that will supply objects such as `HttpCache`, `HttpLoggingInterceptor`, GSON object, and such:

```
@Module
```

```
class NetworkModule {
    @Provides @Singleton
    fun getHttpLoggingInterceptor():HttpLoggingInterceptor=
            HttpLoggingInterceptor().
                    setLevel(HttpLoggingInterceptor.Level.BODY)

    @Provides
    @Singleton
    fun provideHttpCache( @AppContext application: App): Cache {
        val cacheSize = 10 * 1024 * 1024
        val cache = Cache(application.cacheDir, cacheSize.toLong())
        return cache
    }

    @Provides
    @Singleton
    fun provideGson(): Gson {
        val gsonBuilder = GsonBuilder()
gsonBuilder.setFieldNamingPolicy(FieldNamingPolicy.LOWER_CASE_WITH_
UNDERSCORES)
        return gsonBuilder.create()
    }

    @Provides
    @Singleton
    fun provideOkhttpClient(cache: Cache, httpLoggingInterceptor:
HttpLoggingInterceptor): OkHttpClient =
OkHttpClient.Builder().addInterceptor(httpLoggingInterceptor).cache
(cache).build()

    @Provides @Singleton
    fun getRetrofit(okHttpClient: OkHttpClient): Retrofit =
Retrofit.Builder().addCallAdapterFactory(RxJava2CallAdapterFactory.
create())
            .addConverterFactory(GsonConverterFactory.create())
            .client(okHttpClient)
            .baseUrl(AppConstants.INSTAGRAM_BASE_URL)
            .build()

}
```

As you can see, we annotated every method that provides an object through
dependency injection with the @Provides annotation. We have also used
the @Singleton annotation, which means that a singleton object is provided by
the method.

You may notice, we've used other objects to create injectable objects, which we've provided as parameters. Those parameters should be available either from outside or from other injected objects.

2. Now, let's look at an example of the `Dagger` component:

```
@Component(dependencies = arrayOf(ApplicationComponent::class)
        , modules = arrayOf(AddActivityModule::class))
interface AddActivityComponent {
    fun inject(addActivity: AddActivity)
}
```

The component acts as an interface that tells us from which modules (or other components) the dependencies are met. In the preceding example, we created a component that will provide us with the dependency objects from `AddActivityModule` and another `ApplicationComponent` component.

We also defined an inject method, which takes in a parameter (`AddActivity` here) that tells us where the objects will be injected.

3. Once defined, we can inject it into our `AddActivity`, as follows:

```
class AddActivity :
BaseActivity<AddActivityMvpView,AddActivityMvpPresenter>(),AddActiv
ityMvpView {

    @Inject
    lateinit var mPresenter:AddActivityMvpPresenter
    override fun onCreate(savedInstanceState: Bundle?) {
        super.onCreate(savedInstanceState)
        setContentView(R.layout.activity_add)
        DaggerAddActivityComponent.builder()
                .applicationComponent(applicationComponent)
                .build()
                .inject(this)
    }
}
```

As you can see, we use our `AddActivityComponent` (now prefixed with *Dagger*) to inject our `AddActivity` class.

Also, we have marked our dependency objects using the `@Inject` annotation, which means objects will be injected here. We've also added the `lateinit` modifier to prevent us from null checks every time we access it. Adding the `@Inject` annotation means that you want an object there, and dagger will then look into its components and dependencies to provide you that object.

Apart from that module class, you can instantiate the object at constructor level. Let's take a look at the following example:

```
class AddActivityPresenter @Inject constructor(var
mDataManager:DataManager)
```

In the preceding example, adding the `@Inject` annotation to constructor means that the class needs the `DataManager` object before it can be created. Dagger will look into its dependency tree (in its component) and create `AddActivityPresenter`, if present.

Using Butterknife with Kotlin

The Android world has many libraries that require annotation processing. You just annotate the code, and it generates all the code behind the scenes for you, making your life easier. Many libraries such as Butterknife and Dagger2 work in similar ways. In this recipe, we will learn how to use Butterknife with Kotlin. For those who aren't familiar with Butterknife, it's a library that binds a view to a field without needing the `findViewById` calls. It's a household name in the Android development world. In Kotlin, the Kotlin Android Extension does almost the same work and is bundled along with Kotlin. However, if you are migrating your Java code where you've used Butterknife, this recipe will help you.

Getting ready

We will be using Android Studio 3.0 for coding purposes.

How to do it...

To include Butterknife in your project, follow the given steps:

1. To start with, add the following lines to your `build.gradle` file; also, you need to add the `kotlin-kapt` plugin and replace `annotationProcessor` with `kapt`. `kapt` is the Java equivalent of `annotationProcessor`, so wherever you used `annotationProcessor`, you need to replace it with `kapt`:

```
apply plugin: 'kotlin-kapt'
dependencies {  ...
    compile "com.jakewharton:butterknife:$butterknife-version"
    kapt "com.jakewharton:butterknife-compiler:$butterknife-
version" }
```

2. In Java, we used the Butterknife library, as shown:

```
@BindView(R.id.headline) TextView headline;
```

In Kotlin, we can do it as follows:

```
@BindView(R.id.headline) lateinit var headline: TextView
```

Note that we've used the `lateinit` modifier, which will save us from declaring it nullable. We can also implement click listeners, as illustrated:

```
@OnClick(R.id.button)
internal fun sayHello() {
    Toast.makeText(this, "Hello, World!", LENGTH_SHORT).show()
}
```

There's more...

It's important to understand the working of annotation processors. They basically act as a hook for the compiler to analyze the source code for defined annotations, and handle them by producing compiler errors, warnings, or extra code in their place. This makes writing apps faster, because you just have to annotate and the compiler will generate all the necessary code for you behind the scenes. Dagger 2 is a popular library that works this way.

11
Networking and Concurrency

The following recipes will be covered in this chapter:

- How to fetch data over network

- How to create data class

- How to copy data class with modifications

- How to parse JSON data from network to data class

- How to download a file in Kotlin

- How to use RxJava and Retrofit with Kotlin

- How to make an endless list using RecyclerView

- How to use Anko to run background tasks with Kotlin in Android

- How to use coroutines to achieve multithreading

Introduction

You will probably find it hard to get an app that doesn't communicate over the network. Communication with the internet is used in almost all the apps, be it a file-sharing app, streaming app, social network apps, or something else, the list goes on and on. There are many variables that you need to think about when you want to add network communication features to your Android apps. For example, you can't run it on the main thread, and network requests are always performed on background threads. Apart from that, you also need to detect when the network request fails, so as to give feedback to the user about what went wrong. In this chapter, we will address how to efficiently make network requests in Kotlin.

How to fetch data over network

Making a network request in Android is very cumbersome unless you use any third-party library. For example, let's take a look at how network requests in Android used to be:

```
try {
    URL url = new URL("<api call>");

    urlConnection = (HttpURLConnection) url.openConnection();
    urlConnection.setRequestMethod("GET");
    urlConnection.connect();

    InputStream inputStream = urlConnection.getInputStream();
    StringBuffer buffer = new StringBuffer();
    if (inputStream == null) {
        // Nothing to do.
        return null;
    }
    reader = new BufferedReader(new InputStreamReader(inputStream));

    String line;
    while ((line = reader.readLine()) != null) {
        buffer.append(line + "\n");
    }

    if (buffer.length() == 0) {
        return null;
    }
    result = buffer.toString();
} catch (IOException e) {
    Log.e("Request", "Error ", e);
```

```
        return null;
    } finally{
        if (urlConnection != null) {
            urlConnection.disconnect();
        }
        if (reader != null) {
            try {
                reader.close();
            } catch (final IOException e) {
                Log.e("Request", "Error closing stream", e);
            }
        }
    }
}
```

Of course, the preceding code is ugly. Kotlin eases out our pain to make a network request. In this recipe, we will learn how to make network requests in Kotlin.

Getting ready

We will be using Android Studio 3.0. Ensure that you have its latest version.

How to do it...

Let's take look at the following steps required to make network requests in Kotlin:

1. Remember the huge pile of code we saw at the start of this recipe, just for performing a network request? All that can be replaced by just one line of Kotlin code. Let's take a look at the following code:

    ```
    var response= URL("<url>").readText()
    ```

    This will just return the `response` fetched from the network request that you made. You just need to provide your URL as the parameter.

    While using this on Android, ensure that you have pushed this task in the background, or else you will get a `NetworkOnMainThread` exception.

2. The code that we want to execute asynchronously is wrapped under the `doAsync` block. Wrapping the code inside an async task is also very simple. Let's take a look at the following code:

    ```
    doAsync {
        val result=
    ```

```
URL("https://api.instagram.com/319bad89407ffd7082").readText()
    uiThread {
        toast(result)
    }
}
```

We have wrapped the network request into an async block, and then we have a `uiThread` method from where we can touch the UI elements of the app.

3. The `uiThread` method is provided by Anko library, which you can include in your project by adding these lines in your `build.gradle`:

```
implementation "org.jetbrains.anko:anko:1.0"
```

There's more…

Check out the *How to use Anko to run background tasks with Kotlin in Android* recipe of this chapter to learn more about how to create background tasks in Kotlin.

How to create data class in Kotlin

Are you sick and tired of creating long boilerplate code just for storing data? Do you feel that the following code is too much just to define a `Student` model?:

```
public class Student {
    private String name;
    private String roll_number;
    private int age;

    public String getName() {
        return name;
    }

    public void setName(String name) {
        this.name = name;
    }

    public String getRoll_number() {
        return roll_number;
    }

    public void setRoll_number(String roll_number) {
```

```
        this.roll_number = roll_number;
    }

    public int getAge() {
        return age;
    }

    public void setAge(int age) {
        this.age = age;
    }
    @Override
    public int hashCode() {
        return super.hashCode();
    }

    @Override
    public String toString() {
        return super.toString();
    }
}
```

If you agree, then Kotlin's data class is just for you. So let's dive into it in this recipe and get to know it more.

Getting ready

We will be using IntelliJ IDEA to write our code. You can use any IDE that is capable of executing Kotlin code.

How to do it...

In every real-world project, you create classes that don't have any use except for storing data, like in the case of the `Student` class we described earlier. The number of these types of classes can be way too high in a complex project having many roles and models. This results in a lot of boilerplate code. Kotlin has a great solution for it:

1. The code mentioned at the start of the recipe can be reduced to just one line:

   ```
   data class Student(var name:String,var roll_number:String,var
   age:Int)
   ```

   That's it!

2. Now, let's try to use the data class we just created:

```
fun main(args: Array<String>) {
    val student=Student("Aanand","2013001",21)
    println("Student: name- ${student.name},
roll_number:${student.roll_number}, age:${student.age}")
}
```

```
//Output: Student: name- Aanand, roll_number:2013001, age:21
```

As you can see, we didn't need any getter setter, which saved us a lot of boilerplate code. The getter setters are already included in the Kotlin property.

3. Let's check the `toString()` method (which we haven't even defined):

```
println("${student.toString()}")
```

```
//Output: Student(name=Aanand, roll_number=2013001, age=21)
```

This is better than what you'd get from Java's `toString()` method.

4. Data class also offers a lot of flexibility. For example, if you don't want setter for a property, you can make the property `val`. This will make the property read-only:

```
data class Student(val name:String,val roll_number:String,var
age:Int)
```

5. One of the cool things that you can do with data class is that you can destructure the object to obtain the property. Check out the following code to understand more:

```
fun main(args: Array<String>) {
    val student= Student("Aanand", "2013001", 21)
    val (name, roll_number,age)=student
    println("Student: name- $name, roll_number:$roll_number,
age:$age")
}
```

```
//Output: Student: name- Aanand, roll_number:2013001, age:21
```

6. You can also have default values of property in the class. Let's take a look at the next example:

```
data class Student(val name:String="Aanand",val
roll_number:String,var age:Int)
var studentA= Student(roll_number =  "2013001", age = 21)
println(studentA.toString())

//Output: Student(name=Aanand, roll_number=2013001, age=21)
```

There's more...

Data class has a few restrictions. According to Kotlin documentation, those are as follows:

- The primary constructor needs to have at least one parameter
- All primary constructor parameters need to be marked as val or var
- Data classes cannot be abstract, open, sealed, or inner
- Data classes may not extend other classes (but may implement interfaces)

How to copy data class with modifications

In the last recipe, we learned how to use data class and how it reduces a lot of boilerplate code. In this recipe, we will see how data class makes it easy to copy another data class, even if you have to modify the property.

A brute-force mechanism to copy a data class can be to just create a data class by duplicating all the properties, but using the copy method will make it much easier.

Getting ready

We will be using IntelliJ IDEA to write our code. You can use any IDE that is capable of executing Kotlin code.

How to do it...

We will be using the `copy` method, which takes in named arguments and creates a copy of the object with changed values of named arguments. Let's look at an example:

```
data class Student(val name:String,val roll_number:String,var age:Int)
fun main(args: Array<String>) {
    var studentA= Student("Aanand Roy", "2013001", 21)
    var olderStudentA=studentA.copy(age = 25)
    println(olderStudentA.toString())
}

//Output: Student(name=Aanand Roy, roll_number=2013001, age=25)
```

There's more...

People usually get confused between the `copy()` and `apply()` functions:

- `apply()`: It accepts a function and sets its scope to that of the object on which it has been invoked. It is a transformation function that can also be used to evaluate complex logic before returning. At the end, it just returns the same object with changes (if performed).
- `copy()`: The `apply` function is not thread-safe and mutates the object. The `copy()` function, on the other hand, returns a new object (without modifying the original object).

How to parse JSON data from network to data class

JSON is one of the most widely used formats of response. Usually, the APIs provide outputs in the form of JSON response and in Android development also they are used widely as we communicate with network. Parsing JSON response to a data class helps us work with them as a Java object. You can also parse it using JSONObject, but it results in dirty code. In this recipe, we will learn how to parse JSON data into data class. We are using data class because they are preferred when the sole purpose of class is to save data. So let's get started!

Getting ready

We will be using Android Studio 3.0; ensure that you have its latest version. We will be using the GSON library, an open source library by Google for parsing the JSON response. GSON is very easy to use and is one of the most popular JSON parsing libraries out there. To include GSON in your project, just add the following lines to your `build.gradle` file:

```
compile 'com.google.code.gson:gson:2.8.2'
```

How to do it...

Follow these steps to understand how to parse JSON data from network:

1. Usually, we get a JSON response after making a network request, so for simplifying, we will assume that we get the given JSON response after making some network request:

```
{
 "data": [{
        "id": "17867282641151111",
        "from": {
            "id": "1391934316",
            "username": "aanandshekharroy",
            "full_name": "Aanand Shekhar Roy",
            "profile_picture":
"https://scontent.cdninstagram.com/t51.2885-19/10475071_60579025952
7941_865730435_a.jpg"
            },
        "text": "Testing api",
        "created_time": "1501571384"
    }, {
        "id": "17892289033060177",
        "from": {
            "id": "1391934316",
            "username": "aanandshekharroy",
            "full_name": "Aanand Shekhar Roy",
            "profile_picture":
"https://scontent.cdninstagram.com/t51.2885-19/10475071_60579025952
7941_865730435_a.jpg"
 },
            "text": "My second test",
            "created_time": "1501571390"
        }],
        "meta": {
        "code": 200
```

```
        }
    }
```

IntelliJ IDEA provides a plugin that can help convert JSON response into Kotlin object. We will be using the `RoboPojoGenerator` plugin. Carry out the following steps to install it:

2. Go to **Settings** | **Plugins**:

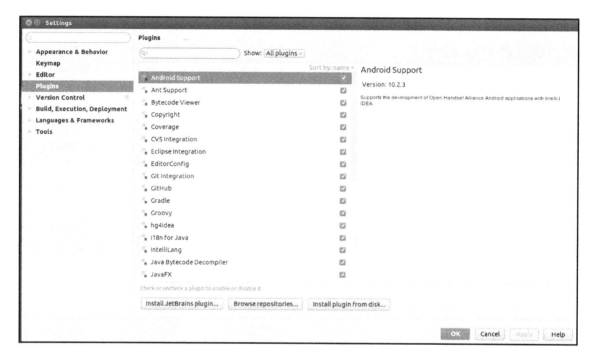

3. Click on **Install JetBrains plugin**; it will open a dialog. Search Robopojo in it, and you will see a **RoboPOJOGenerator** plugin. Click on **Install** and restart the Android Studio.

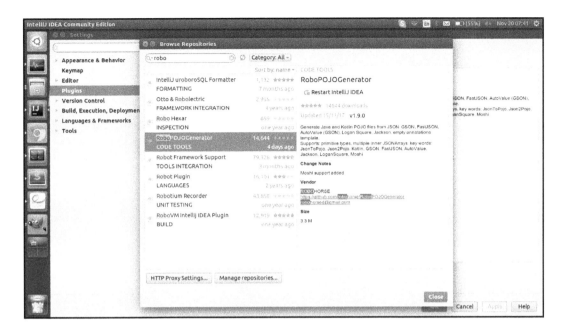

4. Once you've done that, to generate classes based on JSON response, first create an empty package where you would want to keep the classes. I have created it by the name `InstagramCommentsResponse` (because we have used Instagram API for fetching latest comments).

5. Now, right-click on the package, and select **New** | **Generate POJO from JSON**. You will then see a dialog by the RoboPOJO generator, where you need to paste your JSON response. After you've done that, check the **Kotlin** and **Gson** box and click on **Generate**.

6. Now you will see a bunch of classes created inside that package, as illustrated:

7. Let's take a look at these classes. The first class is the outer holder of JSON response:

```
@Generated("com.robohorse.robopojogenerator")
data class Response(
```

```
    @field:SerializedName("data")
    val data: List<DataItem?>? = null,

    @field:SerializedName("meta")
    val meta: Meta? = null
)

// DataItem -  Class that will hold comments
@Generated("com.robohorse.robopojogenerator")
data class DataItem(

    @field:SerializedName("created_time")
    val createdTime: String? = null,

    @field:SerializedName("from")
    val from: From? = null,

    @field:SerializedName("id")
    val id: String? = null,

    @field:SerializedName("text")
    val text: String? = null
)
```

8. Now, let's try to parse the JSON received from the network call. We will try to get the first comment received and access it as we do in Kotlin:

```
fun main(args:Array<String>){
    var response=
URL("https://api.instagram.com/v1/media/1571595528561539504_5812999
640/comments?access_token=5812999640.42ee6f0.9441d5bd909f40319bad89
407ffd7082").readText()
    var gson= Gson()
    val comments=gson.fromJson(response,Response::class.java)
    println(comments.data?.get(0))
}

//Output: DataItem(createdTime=1501571384,
from=From(fullName=Aanand Shekhar Roy,
profilePicture=https://scontent.cdninstagram.com/t51.2885-19/104750
71_605790259527941_865730435_a.jpg, id=1391934316,
username=aanandshekharroy), id=17867282641151111, text=Testing api)
```

As you can see, we can use it as a plain Kotlin object, without needing to use `JSONObject` to parse it using keys, which makes JSON parsing very easy.

How to download a file in Kotlin

We often need to download files in our Android application. The most basic way of doing this will be opening a URL connection and using `InputStream` to read the content of the file and storing it in a local file using `FileOutputStream`; all this is in a background thread using `AsyncTask`. However, we don't want to reinvent the wheel. There are a lot of libraries out there that handle all this stuff very nicely for us and make our work super easy, helping us create clean code.

We can use **Volley** (`https://developer.android.com/training/volley/index.html`), a networking library by developers at Google, which makes network communication very easy and fast. Another one we can use is **OkHttp** (by *Square*), which is very efficient, and we can use it along with **Retrofit** (for HTTP API).

For this recipe, we will be using a networking library called **Fuel**, which is written in Kotlin.

Getting ready

Create a new Android project and add an activity. Now, add fuel dependencies to your project dependencies by adding the following lines in your `build.gradle` and syncing the project:

```
//Fuel - Networking in Kotlin
compile 'com.github.kittinunf.fuel:fuel:$fuel_version' //for JVM
```

Here, `$fuel_version` is the latest version of fuel library.

How to do it...

Follow these steps to download a file in Kotlin:

1. Let's start with a button in our view with an `onClickListener` attached to it. I am also adding a `progressBar` to the view to be able to see the progress of our download. This is my view:

```xml
<?xml version="1.0" encoding="utf-8"?>
<android.support.constraint.ConstraintLayout
xmlns:android="http://schemas.android.com/apk/res/android"
    xmlns:app="http://schemas.android.com/apk/res-auto"
    xmlns:tools="http://schemas.android.com/tools"
    android:layout_width="match_parent"
    android:layout_height="match_parent"
    app:layout_behavior="@string/appbar_scrolling_view_behavior"
tools:context="android.my_company.com.helloworldapp.DownloadFileActivity"
    tools:showIn="@layout/activity_download_file">

    <Button
        android:id="@+id/btn_download_file"
        android:layout_width="wrap_content"
        android:layout_height="wrap_content"
        android:layout_marginBottom="32dp"
        android:layout_marginTop="32dp"
        android:text="@string/download_file"
        app:layout_constraintBottom_toBottomOf="parent"
        app:layout_constraintLeft_toLeftOf="parent"
        app:layout_constraintRight_toRightOf="parent"
        app:layout_constraintTop_toTopOf="parent" />

    <ProgressBar
        android:id="@+id/progressBar"
        style="?android:attr/progressBarStyleHorizontal"
        android:layout_width="0dp"
        android:layout_height="wrap_content"
        app:layout_constraintLeft_toLeftOf="parent"
        app:layout_constraintRight_toRightOf="parent"
        app:layout_constraintTop_toTopOf="parent"
        android:progress="0"/>
</android.support.constraint.ConstraintLayout>
```

2. Let's start by downloading a temporary file. We will be using `https://httpbin.org/` for mocking download file API. The following is the code for downloading a temporary file:

```
class DownloadFileActivity : AppCompatActivity() {

    override fun onCreate(savedInstanceState: Bundle?) {
        super.onCreate(savedInstanceState)
        setContentView(R.layout.activity_download_file)
        setSupportActionBar(toolbar)
        Log.d("ya", filesDir.absolutePath + " " +
filesDir.canonicalPath)

        btn_download_file.onClick {
            progressBar.progress = 0
Fuel.download("http://httpbin.org/bytes/32768").destination {
response, url ->
                File.createTempFile("abcd", ".tmp")
            }.progress { readBytes, totalBytes ->
                val progress = readBytes.toFloat() /
totalBytes.toFloat()
                Log.d("progress", progress.toString())
                progressBar.progress = progress.toInt()*100
            }.response { req, res, result ->
                Log.d("status result",
result.component1().toString())
                Log.d("status res", res.responseMessage)
                Log.d("status req", req.url.toString())
            }

        }
    }

}
```

This is how our UI looks:

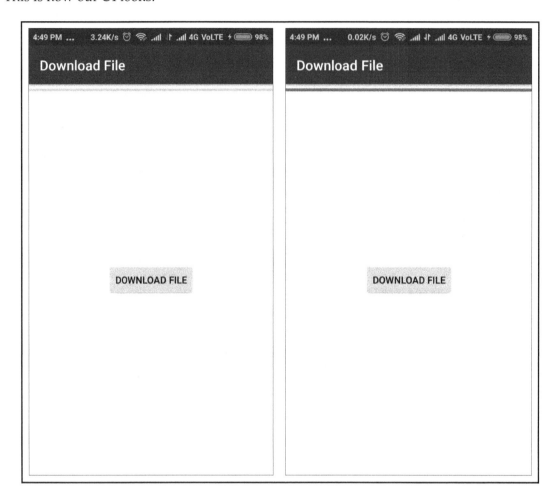

Now, as we mentioned earlier, we will be using the `Fuel` library to download the file; here's how the code looks:

```
btn_download_file.onClick {
            progressBar.progress = 0
            Fuel.download("http://httpbin.org/bytes/32768").destination {
response, url ->
                File(filesDir , "abcd.txt")
            }.progress { readBytes, totalBytes ->
                val progress = readBytes.toFloat() / totalBytes.toFloat()
                Log.d("ya", progress.toString())
```

```
                progressBar.progress = progress.toInt()*100
            }.response { req, res, result ->
                Log.d("status result", result.component1().toString())
                Log.d("status res", res.responseMessage)
                Log.d("status req", req.url.toString())
            }
        }
```

I suggest that you play around with downloading files in Fuel, with and without progress, to get a better grasp of this. You can read all the other functionalities that Fuel provides at `https://github.com/kittinunf/Fuel`.

Also, try to use Volley and other networking libraries in Android to get an understanding of what are the differentiating points and use cases of each.

How to use RxJava and Retrofit with Kotlin

Retrofit is one of the most widely used networking libraries in Android. It is an open source library created by *Jake Wharton*. RxJava is the open source implementation of ReactiveX in Java. RxJava is a great way to do reactive-programming or event-driven programming. This recipe won't teach you about reactive programming (`https://en.wikipedia.org/wiki/Reactive_programming`), so if you aren't comfortable with it, you can learn about it through the documentation (`https://github.com/ReactiveX/RxJava`). Rather, in this recipe, you will learn how to use Retrofit and RxJava together.

Getting ready

We will be using Android Studio 3.0. Ensure that you have its latest version. We also need to add the following dependencies:

```
compile "com.squareup.retrofit2:retrofit:$retrofit_version"
compile "com.squareup.retrofit2:adapter-rxjava2:$retrofit_version"
compile "com.squareup.retrofit2:converter-gson:$retrofit_version"
// RxKotlin - Kotlin version of RxJava
compile "io.reactivex.rxjava2:rxkotlin:$rxKotlinVersion"
```

The `adapter-rxjava2` library helps us return observables as response, which can be subscribed by observers.

How to do it...

Using RxJava with Retrofit is quite simple; let's take a look at the following steps:

1. An instance of Retrofit (which is used to communicate with network) will be created as follows:

```
Retrofit.Builder()
        .addCallAdapterFactory(RxJava2CallAdapterFactory.create())
        .addConverterFactory(GsonConverterFactory.create())
        .client(okHttpClient)
        .baseUrl(AppConstants.INSTAGRAM_BASE_URL)
        .build()
```

2. The following is an example of interface where we define all of our Retrofit calls. You may note, we are returning the Observable, and that's the only thing that has changed:

```
interface InstagramApiService {

    @FormUrlEncoded
    @POST("oauth/authorize")
    fun getRedirectCode(@Field("client_id") client_id: String,
                        @Field("redirect_uri") redirect_uri: String,
                        @Field("response_type") response_type:
String): Call<String>


    @FormUrlEncoded
    @POST("oauth/access_token")
    fun getAccessToken(@Field("client_id") client_id: String,
                        @Field("client_secret") client_secret:
String,
                        @Field("redirect_uri") redirect_uri: String,
                        @Field("grant_type") grant_type: String
                        , @Field("code") code: String)
                         :Observable<InstagramLoginResponse>

    @GET("v1/users/{user_id}/media/recent/")
    fun getInstagramPosts(@Path("user_id") user_id: String?,
@Query("access_token") access_token: String?)
:Observable<InstagramPostsResponse>

    @GET("v1/media/{media_id}/comments")
    fun getCommentsForInstagramPost(@Path("media_id") media_id:
String?
```

```
                                          , @Query("access_token")
access_token: String?)
            :Observable<InstagramCommentsResponse>
}
```

3. Then, you need to create an instance of your aforementioned service, as shown:

```
retrofit.create<InstagramApiService>(InstagramApiService::class.java)
```

4. Now you just need to call the method, and you need a subscriber object that will subscribe to it:

```
instagramApiService.getCommentsForInstagramPost(instagramId)
          .subscribeOn(Schedulers.io())
          .observeOn(AndroidSchedulers.mainThread())
          .subscribeBy(onNext={
              response: InstagramCommentsResponse ->
            // Do something with response

          },onError = {
              // Do something with error
          }))
```

5. We have pushed the network call on a separate thread in the background. When the call/task is complete, the result will be observed on the main thread.

How to make an endless list using RecyclerView

What do the Facebook, Instagram, and Twitter feeds have in common? They all have virtually infinite content to show you while you keep on scrolling down and down for more. There's no doubt that this is a great way to engage users on your platform.

In this recipe, we will see how to make an endless list using RecyclerView. There are many use cases of it, for example, social media, e-commerce application, or any content-based apps.

We will create a simple app, which will load a small set of data in the beginning, but once the user scrolls to the bottom of the content, we will fetch another set of data and append to it, giving an illusion of infinite content to the user. So let's get started!

Getting ready

We will be using Android Studio 3.0; ensure that you have its latest version.

You'll also need to include `RecyclerView` in your `build.gradle` file, which you can add as follows:

```
compile 'com.android.support:recyclerview-v7:26.1.0'
```

You can also find the source code in the https://gitlab.com/aanandshekharroy/Anko-examples/ repository by checking out the `6-endless-list-using-recycler-view` branch.

How to do it...

We will be creating a simple app that will show numbers in the list. As you scroll down, the list will keep growing infinitely:

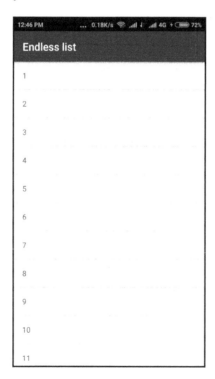

Here's how the list is created:

1. First, we will create an item that will be placed inside the list. The following is the code for that row item `recycler_row.xml`:

```xml
<?xml version="1.0" encoding="utf-8"?>
<LinearLayout
xmlns:android="http://schemas.android.com/apk/res/android"
    android:layout_width="match_parent"
    android:orientation="vertical"
    android:layout_height="wrap_content">

    <TextView
        android:id="@+id/recycler_row_text_view"
        android:layout_width="match_parent"
        android:layout_height="wrap_content"
        android:padding="16dp" />

    <View
        android:layout_width="match_parent"
        android:layout_height="1dp"
        android:layout_alignParentBottom="true"
        android:alpha="0.1"
        android:background="@android:color/black" />
</LinearLayout>
```

2. Next, we will create a `RecyclerView` in the main activity layout file:

```xml
<?xml version="1.0" encoding="utf-8"?>
<android.support.constraint.ConstraintLayout
xmlns:android="http://schemas.android.com/apk/res/android"
    xmlns:app="http://schemas.android.com/apk/res-auto"
    xmlns:tools="http://schemas.android.com/tools"
    android:layout_width="match_parent"
    android:layout_height="match_parent"
    tools:context=".MainActivity">

    <android.support.v7.widget.RecyclerView
        android:id="@+id/recyclerView"
        android:scrollbars="vertical"
        android:layout_width="match_parent"
        android:layout_height="match_parent"/>

</android.support.constraint.ConstraintLayout>
```

3. Now we will create a simple `RecyclerView` adapter:

```
class RecyclerAdapter(val recyclerList: List<Int>) :
RecyclerView.Adapter<RecyclerAdapter.ViewHolder>() {
    override fun onBindViewHolder(viewHolder:
RecyclerAdapter.ViewHolder, position: Int) {
        viewHolder.bind(recyclerList[position])
    }

    override fun onCreateViewHolder(viewGroup: ViewGroup, position:
Int): RecyclerAdapter.ViewHolder {
        val view =
LayoutInflater.from(viewGroup.context).inflate(R.layout.recycler_ro
w, viewGroup, false)
        return ViewHolder(view)
    }

    override fun getItemCount(): Int {
        return recyclerList.count()
    }

    class ViewHolder(itemView: View) :
RecyclerView.ViewHolder(itemView) {
        val itemTextView :TextView=
itemView.findViewById(R.id.recycler_row_text_view)

        fun bind(recyclerItemText: Int) {
            itemTextView.text = recyclerItemText.toString()
        }
    }
}
```

4. Now, let's create a simple function, which when called, will append 30 data items to the list. This is essentially what is done in apps. Once the user reaches the bottom of the list, a network call is made, which appends data to the previous list:

```
fun updateDataList(dataList: MutableList<Int>) : List<Int> {
    kotlin.repeat(30) {
        dataList.add(dataList.size + 1)
    }
    return dataList
}
```

5. Now, let's set up the recycler view in the activity:

```
class MainActivity : AppCompatActivity() {
    val dataList = mutableListOf<Int>()

    override fun onCreate(savedInstanceState: Bundle?) {
        super.onCreate(savedInstanceState)
        setContentView(R.layout.activity_main)

        val layoutManager = LinearLayoutManager(this)
        val adapter = RecyclerAdapter(recyclerList =
updateDataList(dataList))

        recyclerView.layoutManager = layoutManager
        recyclerView.adapter = adapter
        recyclerView.addOnScrollListener(object :
RecyclerView.OnScrollListener() {
            override fun onScrolled(recyclerView: RecyclerView, dx:
Int, dy: Int) {
                super.onScrolled(recyclerView, dx, dy)
                if (!recyclerView.canScrollVertically(1)) {
                    onScrolledToBottom();
                }
            }
        fun onScrolledToBottom() {
            val initialSize = dataList.size
            updateDataList(dataList)
            val updatedSize = dataList.size
adapter.notifyItemRangeInserted(initialSize,updatedSize)
        }
})
}
```

How it works...

Since we have to intercept when the user has reached the bottom of the list, we have added a `ScrollListener`. We have overridden the `onScroll` method, and we are using the `canScrollVertically` method. The `canScrollVertically` method was added in API level 14, and it is used to check whether this view can be scrolled vertically in a certain direction. It takes an integer argument (negative to check scrolling up and positive to check scrolling down) and returns a boolean (`true` if possible, `false` if not). In our example, we have supplied a positive integer, which will return `true` if the view can be scrolled down and false if it can't be. If it can't be further scrolled down (meaning that the data is exhausted), we will add data to the list and update the list by calling the `notifyItemRangeInserted` method of the adapter.

How to use Anko to run background tasks with Kotlin in Android

Anko is a library created by the JetBrains team, which makes Android development quite easy with the help of many helper functions that abstract a lot of complexity and provides you easy-to-use methods. One such thing is to deal with background tasks. Using Anko, we can work with background tasks very easily. In this recipe, we will learn how to work with background tasks using Anko.

Getting ready

We will be using Android Studio 3.0 for coding purposes; ensure that you have its latest version. You need to add Anko to your `build.gradle` file, as shown:

```
implementation "org.jetbrains.anko:anko:$anko_version"
```

How to do it...

Doing a task in the background is very easy in Kotlin. Let's take a look at the next example. In this example, we will make a network request (which is required to do in the background or else you will get a `NetworkOnMainThread` exception); once the network request is complete, we will show the **Success** message using toast. Since we can't touch the UI element from a background thread, we need to come to UI thread in order to do it. We will use the `uiThread` method provided by Anko, which will be called once the background task is over:

```
class MainActivity : AppCompatActivity() {
    override fun onCreate(savedInstanceState: Bundle?) {
        super.onCreate(savedInstanceState)
        setContentView(R.layout.activity_main)
        doAsync {
            val result=
URL("https://api.instagram.com/v1/media/1571595528561539504_5812999640/comm
ents?access_token=5812999640.42ee6f0.9441d5bd909f40319bad89407ffd7082").rea
dText()
            uiThread {
                toast(result)
            }
        }
    }
}
```

As you can see, `URL().readText()` is a long processing task, hence we've put it in the background task.

How it works...

You must have used async tasks for executing tasks in the background, but it wasn't a very efficient way to do it. It has its problem of handling them when the screen rotates, because it doesn't pay attention to the activity's lifecycle.

The `doAsync` method was called by an activity, and if the activity is destroying, the `uiThread` will not be executed. This ensures that you get to touch the UI only till it is present.

How to use coroutines to achieve multithreading

Coroutines are a great language feature in Kotlin. Here's an apt definition of coroutines according to the documentation:

> *"Coroutines are a new way of writing asynchronous, non-blocking code (and much more)."*

It's not just the ease of use, it's much more powerful than threads, especially in the case of a mobile environment where even milliseconds of performance gain is appreciated. Spawning multiple threads can cause performance issues, which isn't the case with coroutines since there can be thousands of those running without much drop in performance levels.

The following is what the official documentation of Kotlin says:

> *"One can think of a coroutine as a lightweight thread. Like threads, coroutines can run in parallel, wait for each other, and communicate. The biggest difference is that coroutines are very cheap, almost free; we can create thousands of them, and pay very little in terms of performance. True threads, on the other hand, are expensive to start and keep around. A thousand threads can be a serious challenge for a modern machine."*

This fact makes it very powerful, and Kotlin's team has provided easy syntax to make it easy to use. In this recipe, we will learn how to use coroutines. So let's get started!

Getting ready

We will be using Android Studio 3.0 for coding purposes. Coroutines are provided as a library that abstracts all the complexities and lets the library handle it. You need to add the library to the `build.gradle` file, like this:

```
dependencies {
    ...
    compile "org.jetbrains.kotlinx:kotlinx-coroutines-core:0.19.2"
}
```

This library is published to the Bintray JCenter repository, so you need to add `jcenter()` in your repositories, as shown:

```
repositories {
  jcenter()
}
```

One thing to note is that coroutines are experimental in Kotlin 1.1, so you need to explicitly tell the compiler that you know it and you are game for it. To do so, you need to add the following lines to your `build.gradle` file:

```
apply plugin: 'kotlin'
  kotlin {
      experimental {
      coroutines 'enable'
  }
}
```

Now that everything is set up, you can start using coroutines in your project.

How to do it...

Let's follow the given steps to understand how coroutines work in Kotlin:

1. There are two functions to start the coroutine:
   - `launch{}`
   - `async{}`

2. Let's try to write our first simple coroutine function:

```
class MainActivity : AppCompatActivity() {
    override fun onCreate(savedInstanceState: Bundle?) {
        super.onCreate(savedInstanceState)
        setContentView(R.layout.activity_main)
        launch {
            delay(10000)
            println("Hello")
        }
    }
}
```

The preceding function will print "Hello" in the Android Studio console after a duration of 10 seconds. Note that we have used the `launch` function to start a coroutine, which returns a Job, but it does not carry any resulting value. It starts a new coroutine on a given thread pool (by default, coroutines are run on a shared pool of threads). Threads still exist in a program based on coroutines, but one thread can run on many coroutines, hence we don't need too many threads.

3. The key to coroutines are suspending functions. We can create a suspending function just by adding the `suspend` modifier on the function. Consider this example:

```
suspend fun timeConsumingMethod(arg: String): Boolean {
    //...
}
```

4. Suspend functions are only allowed to be called from a coroutine or another suspend function. If you try to call them from somewhere else, your code won't even compile.

Antonio Leiva explains a suspending function as follows:

> "*...functions that can stop the execution* when they are called and make it continue once it has finished running their own task."

Coroutines needs to have at least one suspending function (in the last example, `delay` was a suspending function).

5. Next, we will see the `async` function. Conceptually, it is quite similar to the launch function, except the fact that `async` returns a deferred—a lightweight non-blocking future that represents a promise to provide a result later (much like Java's `Future`). To get the result from that deferred, you use `.await()` and since a deferred is also a `Job`, you can cancel it if needed. Let's check out an example of `async`.

6. First, we will create two suspending functions and execute them concurrently, then will add the results from both functions:

```
suspend fun longOperationOne(): Int {
    delay(1000L)
    return 10
}

suspend fun longOperationTwo(): Int {
    delay(1000L)
    return 20
}

val one = async { longOperationOne() }
val two = async { longOperationTwo() }
async {
    println("The answer is ${one.await() + two.await()}")
}
```

7. In the preceding example, we got the deferred objects, which is a `Future` object. To get the result out of it, we have used the `await` function. The `await` function is itself a suspending function; that's why we have wrapped it inside an `async` block.

8. A key thing to note is that both the jobs have run asynchronously and concurrently and hence are non-blocking.

9. If you want to run it in a blocking way, you need to use the `runBlocking` method. Here's the same example, but it will block the main thread while it gets the result:

```
val one = async { longOperationOne() }
val two = async { longOperationTwo() }
runBlocking {
    println("The answer is ${one.await() + two.await()}")
}
```

There's more...

Whenever in doubt, whether using thread or coroutines, remember these lines by Roman (an engineer from the JetBrains team):

> *"Coroutines are for asynchronous tasks that **wait** for something most of the time. Threads are for CPU-intensive tasks."*

In Android's context, you always want to update the UI, but you can't do it from background thread. Coroutines have a solution for this. Let's take a look at the next example:

```
launch(UI) {
    val sum = lengthyJobOne.await() +lengthyJobTwo.await()
    myTextView.text = "Sum of results is $sum."
}
```

In the preceding code, not only can we compute two jobs in the background without blocking the main thread, we can also touch the UI thread for updating views.

12
Lambdas and Delegates

The following recipes will be covered in this chapter:

- Click listeners using lambdas
- Using lazy delegate in Kotlin
- Using the observable delegate
- Using vetoable delegate
- Writing your own delegates
- Using the lateinit modifier
- Working with SharedPreferences
- Creating a chain of multiple lets in Kotlin
- Creating global variables

Introduction

In this chapter, we will explore the functional aspects of Kotlin language. Kotlin has functional programming built in using lambdas. Java was lacking this modern language feature up until now, but it has included lambdas in Java 8. However, since most of the Android devices don't support Java 8, Android developers were not able to use this feature. In this chapter, we will go through them and will also learn about delegates. Delegates are a powerful language feature of Kotlin. So let's get started!

Click listeners using lambdas

An onclick listener in Android are one of those things that used to take up a lot of lines, even if the important portion of the code was only one line. Kotlin simplifies Android framework a lot, and one of the best improvements is `onClickListener`. In this recipe, we will see how we can simplify the traditional lengthy click listeners with the help of lambdas.

Getting ready

I'll be using Android Studio 3 to write code. You can get started by creating a new project in Kotlin with a blank activity in Android Studio 3+, as we won't be using any code from other recipes. You also need an intermediate understanding of Android development.

How to do it...

Let's follow the given steps to understand how to use click listener using lambdas:

1. Let's start with creating an activity with some view, such as a button on which we can attach an `onClickListener`. Check out the following XML layout for one possible activity layout:

```xml
<?xml version="1.0" encoding="utf-8"?>
<android.support.design.widget.CoordinatorLayout
    xmlns:android="http://schemas.android.com/apk/res/android"
    xmlns:app="http://schemas.android.com/apk/res-auto"
    xmlns:tools="http://schemas.android.com/tools"
    android:layout_width="match_parent"
    android:layout_height="match_parent">

    <android.support.design.widget.AppBarLayout
        android:layout_width="match_parent"
        android:layout_height="wrap_content"
        android:theme="@style/AppTheme.AppBarOverlay">

        <android.support.v7.widget.Toolbar
            android:id="@+id/toolbar"
            android:layout_width="match_parent"
            android:layout_height="?attr/actionBarSize"
            android:background="?attr/colorPrimary"
            app:popupTheme="@style/AppTheme.PopupOverlay" />

    </android.support.design.widget.AppBarLayout>
```

```xml
<LinearLayout
    xmlns:android="http://schemas.android.com/apk/res/android"
    xmlns:app="http://schemas.android.com/apk/res-auto"
    xmlns:tools="http://schemas.android.com/tools"
    android:layout_width="match_parent"
    android:layout_height="match_parent"
    android:background="@color/white"
    android:orientation="vertical"
    app:layout_behavior="@string/appbar_scrolling_view_behavior">

    <Button
        android:id="@+id/btn1"
        android:layout_width="match_parent"
        android:layout_height="wrap_content"
        android:layout_margin="8dp"
        android:text="@string/button1"/>

    <Button
        android:id="@+id/btn2"
        android:layout_width="match_parent"
        android:layout_height="wrap_content"
        android:layout_margin="8dp"
        android:text="@string/button2"/>

    <Button
        android:id="@+id/btn3"
        android:layout_width="match_parent"
        android:layout_height="wrap_content"
        android:layout_margin="8dp"
        android:text="@string/button3"/>

</LinearLayout>

</android.support.design.widget.CoordinatorLayout>
```

2. The following is how our layout looks, with three buttons on which we need to attach `onClickListener`:

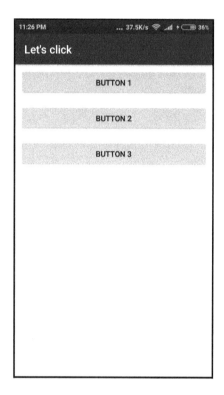

3. Now, let's see the code when we attach `onClickListener` to the three buttons in Java:

```
public class HelloWorldActivity extends AppCompatActivity {
    protected void onCreate(Bundle savedInstanceState) {
        super.onCreate(savedInstanceState);
        setContentView(R.layout.activity_hello_world);
        final Button btn1 = (Button) findViewById(R.id.btn1);
        final Button btn2 = (Button) findViewById(R.id.btn2);
        final Button btn3 = (Button) findViewById(R.id.btn3);
        final Toolbar toolbar = (Toolbar)
findViewById(R.id.toolbar);
        setSupportActionBar(toolbar);
        getSupportActionBar().setTitle("Let's click");
        btn1.setOnClickListener(new View.OnClickListener() {
            public void onClick(View v) {
                Toast.makeText(HelloWorldActivity.this, "Button 1
```

```
has been clicked by you! One", Toast.LENGTH_SHORT).show();
            }
        });
        btn2.setOnClickListener(new View.OnClickListener() {
            public void onClick(View v) {
                Toast.makeText(HelloWorldActivity.this, "Button 2
has been clicked by you! Two.", Toast.LENGTH_SHORT).show();
            }
        });
        btn3.setOnClickListener(new View.OnClickListener() {
            public void onClick(View v) {
                Toast.makeText(HelloWorldActivity.this, "Button 3
has been clicked by you! Three", Toast.LENGTH_SHORT).show();
            }
        });
    }
}
```

4. Did you note the amount of code we had to write just for attaching click listeners that show toasts? All this code for just three onclick listeners; now, let's see the difference it makes to write the same code in Kotlin:

```
class HelloWorldActivity2 : AppCompatActivity() {
    override fun onCreate(savedInstanceState: Bundle?) {
        super.onCreate(savedInstanceState)
        setContentView(R.layout.activity_main)
        setSupportActionBar(toolbar)
        supportActionBar?.title = "Let's click"
        btn1.setOnClickListener(object : View.OnClickListener {
            override fun onClick(v: View?) {
                toast("Button 1 has been clicked by you! One")
            }
        })
        btn2.setOnClickListener(object : View.OnClickListener {
            override fun onClick(v: View?) {
                toast("Button 2 has been clicked by you! Two")
            }
        })
        btn3.setOnClickListener(object : View.OnClickListener {
            override fun onClick(v: View?) {
                toast("Button 3 has been clicked by you! Three")
            }
        })
    }
}
```

5. The code is certainly lesser and cleaner, thanks to Kotlin **synthetic properties** and **Anko** for toast helper function. Now, let's try to use **lambda** and see what difference it makes:

```
class HelloWorldActivity2 : AppCompatActivity() {

    override fun onCreate(savedInstanceState: Bundle?) {
        super.onCreate(savedInstanceState)
        setContentView(R.layout.activity_main)
        setSupportActionBar(toolbar)
        supportActionBar?.title = "Let's click"
        btn1.setOnClickListener({ toast("Button 1 has been clicked
by you! One") })
        btn2.setOnClickListener({ toast("Button 2 has been clicked
by you! Two") })
        btn3.setOnClickListener({ toast("Button 3 has been clicked
by you! Three") })
    }
}
```

Whoa! All hail to the power of lambdas. Note how much code has been reduced, and it looks clean and more readable. This a large amount of boilerplate code reduction, which saves us both time and effort.

How it works...

Lambda functions are functions that are not declared but passed as an expression. In Kotlin, if a function receives an interface, we can replace it with a lambda. For example, the `setOnClickListener` function receives `View.OnClickListener`, so we can use a lambda:

```
fun setOnClickListener(listener: (View) -> Unit)
someView.setOnClickListener({ view -> doSomething() })
```

Also, if there are no parameters to be passed to the lambda function, we can omit the arrow, and if the last parameter being passed is a function, we can move it outside the parentheses:

```
someView.setOnClickListener() { doSomething() }
```

Then, if the lambda function being passed is actually the only parameter, you can omit the parentheses completely:

```
button.setOnClickListener { doSomething() }
```

There's more...

We can use **Anko**, a library by Kotlin to reduce the code even further. Anko provides an `onClick()` method that accepts a lambda that gets executed `onClick` event:

```
class HelloWorldActivity2 : AppCompatActivity() {
    override fun onCreate(savedInstanceState: Bundle?) {
        super.onCreate(savedInstanceState)
        setContentView(R.layout.activity_main)
        setSupportActionBar(toolbar)
        supportActionBar?.title = "Let's click"
        btn1.onClick { toast("Button 1 has been clicked by you! One") }
        btn2.onClick { toast("Button 2 has been clicked by you! Two") }
        btn3.onClick { toast("Button 3 has been clicked by you! Three") }
    }
}
```

Using lazy delegate in Kotlin

The lazy construct is basically used for lazy initialization of properties, which is especially helpful when the object being initialized is a heavy object (takes time to initialize). Instantiating heavy objects at startup can cause visible performance drop in mobile user experience. Lazy initialization can solve our problem. In this recipe, we will learn how to use Kotlin's lazy delegate, so let's get started!

Getting ready

We will be using Android Studio 3.0 for coding, ensure that you have the latest version downloaded.

How to do it...

In the following steps, we will learn how to use the lazy delegate in Kotlin:

1. First, let's see how to create a property through lazy initialization. The syntax is as follows:

```
val / var <property name>: <Type> by <delegate>
```

2. While creating a lazy delegate, we use `by lazy`, as shown:

```
class MainActivity : AppCompatActivity() {
    private val textView : TextView by lazy {
        findViewById<TextView>(R.id.textView) as TextView
    }
    override fun onCreate(savedInstanceState: Bundle?) {
        super.onCreate(savedInstanceState)
        setContentView(R.layout.activity_main)
        textView.text="ABC"
    }
}
```

The lazy delegate initializes the object on its first access and stores the value, which is then returned for subsequent accesses.

How it works...

The delegation of a property looks like this:

```
class Delegate {
    operator fun getValue(
            thisRef: Any?,
            property: KProperty<*>
    ): String {
        // return value
    }
    operator fun setValue(
            thisRef: Any?,
            property: KProperty<*>, value: String
    ) {
        // assign
    }
}
```

The read operation calls the `getValue` method, and the write operation calls `setValue`.

There are three modes of evaluation of lazy properties:

- `LazyThreadSafetyMode.SYNCHRONIZED`: Initialization occurs only on one thread. The rest of the threads see the cached value. It is also the default mode of initialization.

- `LazyThreadSafetyMode.PUBLICATION`: Used when synchronization of initialization delegate is not required. It can then be called from multiple threads at a time, and initialization can be performed on every thread. However, if initialization is done by one thread, it will be returned without performing initialization.
- `LazyThreadSafetyMode.NONE`: No locks are used to synchronize initialization and hence less overhead costs.

Using the observable delegate

Previously, we saw how to work with delegated properties. In this recipe, we will learn how to work with the observable delegate. This delegate helps us observe any changes to the property. So let's get started.

Getting ready

We will be using IntelliJ IDEA for writing code. You can use any IDE where you are able to execute Kotlin code.

How to do it...

The observable delegates take in a default value and a construct where we have old and new values. Let's take a look at the next example:

```
fun main(args: Array<String>) {
    var a:String by Delegates.observable("",{_,oldValue,newValue ->
        println("old value: $oldValue, new value: $newValue ")
    })
    a="a"
    a="b"
>}
//Output:old value: , new value: a
        old value: a, new value: b
```

In the preceding example, we have provided the initial value as an empty string. The construct will be executed every time we try to update the value of the a property. We have changed the value of a two times and hence we are seeing two print statements.

There's more…

The observable delegate can be especially useful in the case of `RecyclerView`, because we can use `DiffUtils` to update just the items that are updated, rather than replacing the whole list with a new one. For more information, refer to the recipe in `Chapter 4`, *Creating RecyclerView Adapter in Kotlin*.

Using vetoable delegate

Vetoable delegate is quite similar to the observable delegate, with the only difference of vetoing the change. In the observable delegate, we could get hold of new and old values whenever the observable property was changed. Let's take a look at the definition provided in Kotlin's documentation:

> *"Returns a property delegate for a read/write property that calls a specified callback function when changed, allowing the callback to veto the modification."*

Getting ready

I'll be using IntelliJ IDEA for coding purposes. You can use any IDE capable of executing Kotlin code.

How to do it…

Let's now look at the given steps to understand the `vetoable` modifier:

1. Let's take a quick look at an implementation of the `vetoable` delegated property:

```
fun main(args: Array<String>) {
    var student:Student by
Delegates.vetoable(Student(10),{property, oldValue, newValue ->
        if(newValue.age>25){
            println("Age can't be greater than 25")
            return@vetoable false
```

```
        }
        true
    })
    student=Student(26)
}
class Student(var age:Int)

//Output: Age can't be greater than 25
```

2. As you can see, the modification is "vetoed" by the vetoable delegate since the age can't be greater than 25. The new object will only be assigned if the age is less than 25.

How it works...

Let's take a look at the vetoable delegated property declaration:

```
inline fun <T> vetoable(
initialValue: T,
crossinline onChange: (property: KProperty<*>, oldValue: T, newValue: T) ->
Boolean
): ReadWriteProperty<Any?, T> (source)
```

The vetoable() takes an initial value, which could be an empty list, and also an onChange callback, which is called before the change to a property is made. The callback returns true if the change is successful and false if it is vetoed.

There's more...

Vetoable can be especially useful if you use it in the Recyclerview adapter. Generally, you would assign data to the list directly and may call notifyDatasetChanged, but this is highly inefficient, as it will result in loading all the data again. We can use vetoable to check whether the content is the same by matching the old value and new value and can veto modification if it is the same. Also, we can use DiffUtils to just update the data that is changed. DiffUtils was introduced in Android support library 26.01 and later versions, and makes RecyclerView much more efficient.

Writing your own delegates

Delegated properties are one of the best features of Kotlin language. We have already seen observable and vetoable delegates. In this recipe, we will learn how to create our own custom delegate. As a demo example, we will create a delegate property that can only be initialized once; if done again, it should throw an exception. So let's dive into it and see how we can achieve it.

Getting ready

We will be using IntelliJ IDEA for coding purposes. You can use any IDE capable of executing Kotlin code.

How to do it...

Now, let's dive in and learn how to create our own delegates:

1. Let's create a custom delegate named as `SingleInitializationProperty`. This custom delegate property will throw an exception if the variable isn't initialized, and it can only be initialized once. Doing it a second time will throw an exception. Let's take a look at our custom delegate class:

```
class SingleInitializableProperty<T>() : ReadWriteProperty<Any?,
T>{
    private var value: T? = null
    override fun getValue(thisRef: Any?, property: KProperty<*>): T
{
        if(value==null){
            throw IllegalStateException("Variable not initialized")
        }else {
            return value!!
        }
    }
    override fun setValue(thisRef: Any?, property: KProperty<*>,
value: T) {
        if(this.value==null){
            this.value=value
        }else{
            throw IllegalStateException("Cannot be initialized
twice")
        }
    }
```

```
    }
```

2. Now, that we have created a custom delegate, let's try to use it without initializing it in the following way:

```
fun main(args: Array<String>) {
    var a:String by SingleInitializableProperty()
    println(a)
}
```

This is the output:

Output: **Exception in thread "main" java.lang.IllegalStateException: Variable not initialized**

3. Let's see another example; this time, we will first initialize it, then access it, and then again try to initialize it:

```
fun main(args: Array<String>) {
    var a:String by SingleInitializableProperty()
    a="first"
    println(a)
    a="second"
}
```

Here's the output:

Output: **first**
Exception in thread "main" java.lang.IllegalStateException: Cannot be initialized twice

How it works...

As you can see, we have implemented the ReadWriteProperty interface in the delegated property, which basically means our variable will be of the var type. If you want it to be immutable, you can implement the ReadOnlyProperty interface.

The getValue function takes a reference to a class and a property's metadata. The setValue function, in turn, receives a set value. In case of immutable property (val), there will be only one getValue function.

Using the lateinit modifier

Lateinit is an important initialization property, because if you don't want to initialize your variable in constructor, `lazy` and `lateinit` can be employed to do so. In this recipe, we will see how to use the `lateinit` modifier and how it is different from the `lazy` modifier.

Getting ready

I'll be using IntelliJ IDEA for the coding purpose; you can use any IDE that can execute Kotlin code.

How to do it...

Let's follow the given steps to understand how the `lateinit` modifier works:

1. In Java, we could just declare a variable beforehand and initialize it later, but Kotlin requires you to initialize it as soon as you declare it (unless you are using special modifiers). So you can do the following:

   ```
   var student:Student?=null
   ```

   Alternatively, you can do this:

   ```
   val student=Student()
   ```

   Both ways have their drawbacks. The first way will require you to check nullability whenever you use it, and the second way of initializing will make it immutable.

2. To overcome limitations, we can use a `lateinit` modifier, with which we can declare it beforehand and initialize anywhere we want (but before we first access it). This is especially needed when you use dependency injection. Let's see an example from Kotlin documentation that uses the `lateinit` modifier to declare the variable:

   ```
   public class MyTest {
       lateinit var subject: TestSubject
       @SetUp fun setup() {
           subject = TestSubject()
       }
       @Test fun test() {
   ```

```
            subject.method()  // dereference directly
        }
    }
```

3. If you try to access the variable before initializing it, you will get
   `UninitializedPropertyAccessException`. If you are using dependency
   injection, here's how you would use `lateinit` with it:

```
@Inject
 lateinit var mPresenter:EducationMvpPresenter
```

There's more...

Another way of initializing properties is with the `lazy` modifier; `lazy()` is basically a
function that takes a lambda and returns an instance of lazy, which serves as a delegate for
implementing a lazy property. Let's take a look at the next example:

```
public class Student{
    val name: String by lazy {
        "Aanand Shekhar Roy"
    }
}
```

By `lazy` initialization, we postpone initialization until we first use it. The property is
initialized only when we first access it, and the same value is returned for subsequent
accesses. That's why it is mandatory to mark the variable immutable. This can really help us
with initialization of heavy objects, which takes a lot of time. Initializing them `lazily` can
improve our startup time. The only con is that you won't be able to modify it later since it is
a `val` property.

Working with SharedPreferences

SharedPreferences is a persistent way of data storage in Android devices and is mostly
used to save data in key-value pairs, such as the settings of an app. Kotlin makes it easier to
work with shared preference using its unique language construct. In this recipe, we will see
how Kotlin can help us deal with SharedPreferences easily. So let's get started.

Getting ready

We will be using Android Studio 3.0 for this recipe. If you have an older version of Android Studio, either update it to 3.0 or configure Kotlin in it.

How to do it...

To be able to define and use SharedPreferences, we follow particular steps. We will go through each step and implement this together:

1. First, we will create a `Prefs` class that will act as a single entry to read/write from our app's SharedPreferences. This will make it easier to handle all SharedPreferences since they all will be in one place. As we know, shared preference requires context to be present, so we will pass context in the primary constructor. We will also create a single SharedPreferences object that we will use throughout the class:

   ```
   class Prefs (mContext:Context){
       val
   sharedPrefences=mContext.getSharedPreferences("com.ankoexamples.app
   ",Context.MODE_PRIVATE)
       val PREF_USERNAME="pref_username"
   }
   ```

2. For example, we have defined a PREF_USERNAME SharedPreferences; here, we will store the username of the user. Now the fun part begins; remember that Kotlin has a property where we can explicitly define how to get and set the property. We will use the same thing here. Let's take a look at the given code:

   ```
   var username:String
       get() = sharedPrefences.getString(PREF_USERNAME,null)
   set(value)=sharedPrefences.edit().putString(PREF_USERNAME,value).ap
   ply()
   ```

   As you can see, in the setter, we are editing the shared preference and in the getter, we are extracting the value of the shared preference.

3. Now that we have our `Prefs` class ready, we can use it in our activities, fragments, and so on. The best way to do it will be by defining it in the `Application` class and accessing is from many activities or fragments, because then we will not need to create multiple objects of the `Prefs` class. So let's create an `Application` class and a singleton instance of the `Prefs` class:

```
class App:Application() {
    companion object {
        var prefs: Prefs? = null
    }

    override fun onCreate() {
        prefs = Prefs(this)
        super.onCreate()
    }
}
```

We have added and placed our `prefs` variable inside the companion object to be able to use it statically. Also, now that we have placed it inside the `Application` class, we will be dealing only with a single instance of the `prefs` object.

4. We can also use the `lazy` construct to ensure that we create an object only at its first access. Doing so will also help us avoid null checks. Here's how our `App` class will look:

```
val prefs: Prefs by lazy {
    App.prefs!!
}
class App:Application() {
    companion object {
        var prefs: Prefs? = null
    }
    override fun onCreate() {
        prefs = Prefs(this)
        super.onCreate()
    }
}
```

5. Let's now look at an example to add a value to our SharedPreferences:

```
class MainActivity : AppCompatActivity() {

    override fun onCreate(savedInstanceState: Bundle?) {
        super.onCreate(savedInstanceState)
        setContentView(R.layout.activity_main)
```

```
        prefs.username="Aanand"
    }
}
```

6. It looks so simple to work with shared preferences now, as if we are assigning values to a variable. Accessing them is also very easy:

```
Log.d(prefs.username) // Aanand
```

There's more...

As you can see, we have used the `apply()` method to save preferences, which commits the changes in the in-memory SharedPreferences immediately, but also starts an asynchronous commit to the disk; `commit()`, on the other hand, writes to persistent storage synchronously.

Creating a chain of multiple lets in Kotlin

`let` is a pretty useful function provided by Kotlin's `Standard.kt` library. It is basically a scoping function that allows you declare the variable in its scope. Let's take a look at the given code:

```
someVariable.let{
    // someVariable is present as "it"
}
```

However, the best thing is that it can be used to avoid null checks. Earlier, you might have used the following:

```
if(someVariable!=null){
    // do something
}
```

While the preceding code is good, it's not very suited for mutating properties. The alternative is to use `?.let` (`someVariable.?let{}`), which ensures that the code block runs when the variable is not null. However, what if we have multiple if-not-null chains? Let's see how to deal with those cases in this recipe.

Getting ready

We will be using IntelliJ IDEA to write code. You can use any IDE that is capable of executing the Kotlin code.

How to do it...

Follow the mentioned steps to understand how to create a chain of multiple lets:

1. When you have to do multiple null-checks, you can obviously use nested `if-else`, checking null conditions, as in the following code:

```
if(variableA!=null){
    if(variableB!=null){
        if(variableC!=null){
            // do something.
        }
    }
}
```

2. Since we know that the `let` function guarantees that the block will run only when the object is not-null, we need to create a function that will perform the function of `let` but on three variable scenarios. Let's take a look at our function:

```
fun <T1: Any, T2: Any,T3:Any, R: Any> multiLet(p1: T1?, p2:
T2?,p3:T3?, block: (T1, T2,T3)->R?): R? {
    return if (p1 != null && p2 != null &&p3!=null) block(p1,
p2,p3) else null
}
```

3. Now we can use it as illustrated:

```
fun main(args: Array<String>) {
    var variableA="a"
    var variableB="c"
    var variableC="b"
    multiLet(variableA,variableB,variableC){
        _,_,_->
        println("Everything not null")
    }
}

//Output: Everything not null
```

4. In a similar way, it can be employed for two variable scenarios. You might be thinking how to do it in a multi-object scenario, like in the case of a list. Let's create a `whenAllNotNull` function, which will run the block of code only when all the elements of the list are not null:

```
var nonNullList=listOf("a","b","c")
nonNullList.whenAllNotNull {
    println("all not null")
}
fun <T: Any, R: Any> Collection<T?>.whenAllNotNull(block:
(List<T>)->R) {
    if (this.all { it != null }) {
        block(this.filterNotNull())
    }
}
```

```
Output: all not null
```

Creating global variables

In Java, we could create a global variable just by defining the variable at the beginning of class declaration and initializing it afterward. By just declaring it, we could use it as a global variable.

In this recipe, we will learn how to create and use a global variable in Kotin.

Getting ready

I'll be using IntelliJ for coding purposes. You can use any IDE where you can write and execute Kotlin code.

How to do it...

Now, let's look at how to create global variables in Kotlin. There are two ways to do it. Let's look at them one by one:

1. One way to do it is by declaring it under the class declaration. We can use `var` declaration, like this:

```
fun main(args: Array<String>) {
    var student:Student?=null
}
```

However, this approach will result in testing for nullability whenever you use it:

```
println(student?.age)
```

2. To prevent this, you can declare and initialize it using `val`, but that will result in an immutable variable, which might not be the desired behavior.

3. Another way to declare a global variable is by using the `lateinit` modifier. Here's how the preceding code will look:

```
fun main(args: Array<String>) {
    lateinit var student:Student
    student=Student()
    println(student.age)
}
```

4. The `lateinit` modifier is used to first declare the variable, without needing it to be defined either null or immutable. However, we need to initialize it before we use it; otherwise, it will throw an `UninitializedPropertyAccessException`.

 The `lateinit` modifier doesn't work with primitive types.

5. `lateinit` can also be useful when you try to initialize the variables using dependency injection. This way, you avoid null checks when referencing the property inside the body of a class.

13
Testing

The following recipes will be covered in this chapter:

- Unit testing Kotlin code

- Unit testing with Mockito

- Running Instrumentation tests

- Writing JUnit rules in Kotlin (@Rule)

- Acceptance tests using Espresso Kotlin

- Writing assertEquals in Kotlin

Introduction

Testing is a fundamental part of software engineering if you want your code base to be scalable and maintainable. In Android, there are basically two types of testing: one is **unit testing** and the other is **integrated testing**. Unit testing is a type of testing where individual units are tested independently, while integrated testing, which is also sometimes known as instrumentation testing, requires an Android device or an emulator for the tests to run. Since integrated testing requires real devices or an emulator, these tests are often slower to execute. Unit tests are fast because they don't have any such need for real devices or emulators in order to run. Since unit tests are faster and instrumentation tests are slower, it is often thought that a robust test suite should have these tests in the proportion of 80% to 20%. So your code base should consist of 80% unit tests and 20% instrumentation tests.

Unit testing Kotlin code

Unit tests involve, basically, *testing in units*. These tests are often faster to execute because they are executed in the JVM, and hence do not require the dexing, packaging, and installing-on-the-emulator steps, reducing test cycles from minutes to seconds so that you can quickly iterate and refactor your code. Integration tests, on the other hand, require all the aforementioned steps. Apart from testing your code, unit tests also work as a great documentation of the code base. That's why it shouldn't surprise you if you see the names of methods phrased in odd ways—for example, `testIfConfirmationEmailIsSent`.

In this recipe, we will learn how to write unit tests for your Android code.

Getting ready

You'll need Android Studio, as we will learn to write unit tests for Android code, and also because Android Studio provides great support for unit tests. You can also find the source code at `https://gitlab.com/aanandshekharroy/Anko-examples` in the **4-unit-tests** branch.

How to do it...

Follow these steps to understand how to write unit tests for Android code in the Kotlin language:

1. When you create a new Android project in Android Studio, Android Studio will provide support for both unit and Android tests. It provides you with separate directories where you can place your test. Take a look at the following screenshot:

As you can see, **test** is where *unit tests* are placed and **androidTest** is where *Android tests* or *instrumentation tests* are placed.

2. There are two demo tests provided to you already: **ExampleUnitTest** and **ExampleInstrumentedTest**. To run them, just right-click on **ExampleUnitTest** and click on **Run ExampleUnitTest**. After running the tests, you can see the results of the tests in the console, as seen in the following screenshot:

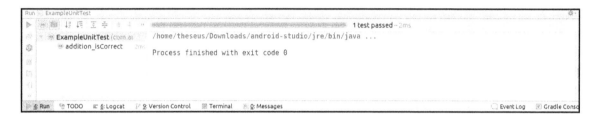

Now, let's try creating our own unit test:

1. We usually have the Utility class in our code, which has methods that can be used by any class, so instead of defining those methods in every class, we define them in the Utility class. So let's create a method, addTwoNumbers, which will take two parameters, a and b, and return a result—a+b:

```
class Utility {
  companion object {
      fun addTwoNumbers(a:Int, b:Int):Int=a+b
  }
}
```

2. In Android Studio, you can create tests directly from the class itself. Just right-click on the class name and click on **Create test**:

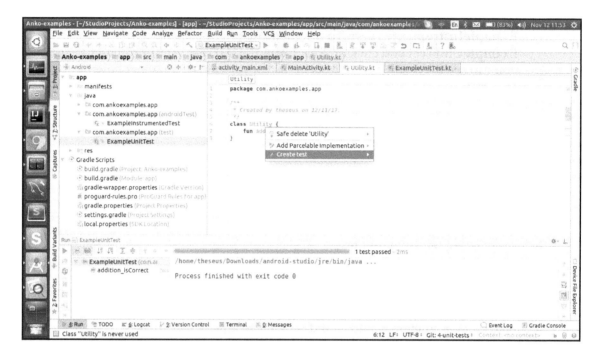

3. After that, you will be shown a dialog on which you can select all the methods for which you want to create a test. It is recommended that you have tests for every method:

4. When you click on **OK**, you will be shown another dialog box that will ask you where to place your tests. In this case, this is a unit test, so we will place it in `app/src/test...`:

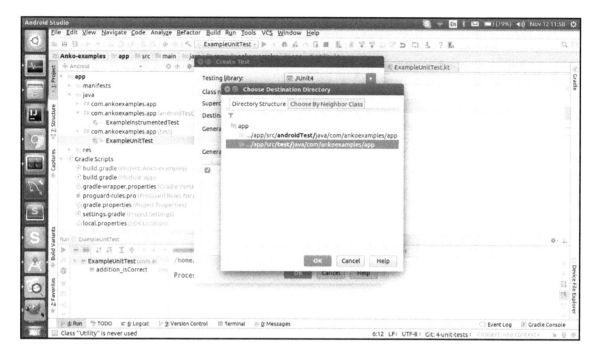

5. When you click on **OK**, Android Studio autogenerates boilerplate code in `UtilityTest.kt`, which looks like this:

```
/**
 * Created by theseus on 12/11/17.
 */
class UtilityTest {
    @Test
    fun addTwoNumbers() {
    }
}
```

6. Now we will add a few `assertEquals` statements, which check the expected value against the results:

```
class UtilityTest {
    @Test
    fun addTwoNumbers() {
```

```
        assertEquals(5,Utility.addTwoNumbers(2,3))
        assertEquals(5,Utility.addTwoNumbers(4,1))
        assertNotEquals(5,Utility.addTwoNumbers(2,5))
    }
}
```

The first argument is the expected value and the second argument is the output of the function. If you run the test, it will pass.

How it works...

When you run the unit tests, they test all the methods that have the `@Test` notation above them. There is also a `@Before` notation, which is placed above a method. The method with the `@Before` notation is run before any other method of the class. This can be helpful when you are setting up objects and variables that might be used later.

One thing to note here is that unit tests cannot use Android SDK components. To use those components, you need instrumentation tests, or use a mocking framework like **Mockito**, which mocks the Android components so that it can be used unit tests. We will cover this in the next recipe.

Unit testing with Mockito

As we discussed in the previous recipe, we cannot use Android components in unit tests. This is why we are able to run them faster, and without any device. If you want to use Android components in your tests, there are two options:

- Write integration tests, which run on your device or emulator.
- Use a mocking framework, such as Mockito, which basically mocks the Android SDK components so that you can use them without any device or emulator, just like any other unit test. The benefit of a mocking framework is that it takes a lot less time to run the tests, as the tests are basically unit tests only. Here's an accurate definition of a mock object by Vogella:

"A mock object is a dummy implementation for an interface or a class in which you define the output of certain method calls. Mock objects are configured to perform a certain behavior during a test. They typical record the interaction with the system and test can validate that."

With that in mind, let's try using Mockito to write unit tests.

Getting ready

You'll need Android Studio, as it provides great support for unit tests, and we will also be learning to write the unit tests for Android code.

You can also find the source code at `https://gitlab.com/aanandshekharroy/Anko-examples` in the **4-unit-tests** branch.

First, you need to add the Mockito dependency to your project. You can do so by adding the following line in your `build.gradle` file at the app level:

```
testImplementation 'org.mockito:mockito-core:2.8.47'
```

Once you have added the dependency, you are good to go ahead.

How to do it...

We usually use Mockito to mock Android classes, but let's test a simple class using Mockito:

1. Here's a small test that tests a `functionUnderTest` function of the `Utility` class:

   ```
   @Test
   fun test_functionUnderTest(){
       val classUnderTest= mock(Utility::class.java)
       classUnderTest.functionUnderTest()
       verify(classUnderTest).functionUnderTest()
   }
   ```

2. In the preceding class, we are calling the `functionUnderTest` method and then verifying whether the method has been called or not. (Yeah!, not a very good use case of a test, but let's try to run this test) When you run it, you'll see an error like this:

   ```
   org.mockito.exceptions.base.MockitoException:
   Cannot mock/spy class com.ankoexamples.app.Utility
   Mockito cannot mock/spy because :
    - final class
   ```

3. The reason for the preceding error is that every class is final by default in Kotlin. You need to *open* them if you want to extend or mock them. However, does that mean that you need to add open modifier to every class that you want to test? That sounds like a bad idea, and it is. There is a hack around this problem. The hack is to manually add the option of mocking the final class. You need to create a file in `test/resources/mockito-extensions folder` called `org.mockito.plugins.MockMaker` and put the following code into that file:

```
mock-maker-inline
```

Now if you run the code, it will pass smoothly.

4. There are many variants of the `verify` method, such as the following:
   - `verify(classUnderTest, never()).functionUnderTest()`, which tests whether it isn't ever called
   - `atLeastOnce()`, `atLeast(2)`, `times(5)`, `atMost(3)`, which can also be used to verify the number of interactions with the method

5. Let's see another Mockito test that mocks `SharedPreferences` (an Android component):

```
@Test
fun testSharedPreference(){
    val sharedPreferences=mock(SharedPreferences::class.java)
    `when`(sharedPreferences.getInt("random_int",-1)).thenReturn(1)
    assertEquals(sharedPreferences.getInt("random_int",-1),1)
}
```

The `when(...).thenReturn(...)` construct keeps an eye on the object, and when the method inside the `when` construct is called, it returns the value under the `thenReturn` construct. Note the `` ` `` surrounding `when`; this is because `when` is a reserved keyword in Kotlin, so we call it with back ticks.

6. You can also return multiple values, which simulates calling a method multiple times:

```
@Test
fun testSharedPreference(){
    val sharedPreferences=mock(SharedPreferences::class.java)
    `when`(sharedPreferences.getInt("random_int",-1)).thenReturn(1).the
nReturn(2)
    assertEquals(sharedPreferences.getInt("random_int",-1),1)
```

```
            assertEquals(sharedPreferences.getInt("random_int",-1),2)
    }
```

In the preceding example, the first call to the `getInt` method will return 1, and the second call will return 2.

There's more...

Let's understand a `spy` object in unit testing.

Spy object

The mocking framework also provides a `spy` method that can be used to wrap the real objects. The calls to the spy objects are delegated to the real object. What's the use of that, you might be thinking. It can check the interactions on a real object, which wasn't possible if the object is not mocked. Let's take a look at the following example:

```
@Test
fun testSpyObject(){
    val list = List(2,init = {-1})
    val spy= spy(list)
    assertEquals(spy.get(0),-1)
    verify(spy).get(0)
}
```

The preceding test will pass.

Note that calling `spy.get(0)` returns -1, which is equal to what you'd get if you had interacted with the real object. Furthermore, you are also able to verify the interaction.

Mockito limitations

Mockito has certain limitations—for example, you cannot mock `static` and `private` methods. This is out of the scope of this book, so to know more about the Mockito limitations, visit `https://github.com/mockito/mockito/wiki/FAQ#what-are-the-limitations-of-mockito`.

Running instrumentation tests

In the preceding recipes, we learned how to run and write unit tests. In this recipe, we will learn how to run instrumentation tests. The integration tests are placed under the **androidTest** directory in your Android project.

Getting ready

Since instrumentation tests require real devices or emulators to run on, ensure that you have one of either of these. We'll be using Android Studio 3.0 for our coding purposes. You can download the source code from `https://gitlab.com/aanandshekharroy/Anko-examples` and switch to the `5-instrumentation-tests` branch. We will also be using *Espresso* for writing instrumentation tests as it is the easiest software to use. Espresso is automatically included in your project when you create a new project.

 Espresso is targeted at developers who believe that automated testing is an integral part of the development lifecycle. While it can be used for black-box testing, Espresso's full power is unlocked by those who are familiar with the code base under test.

How to do it...

In the following steps, you will learn how to run instrumentation tests:

1. Let's create a simple app, which will just have the **Hello World!** text and a button:

2. On clicking the button, the text will change to **Goodbye World!**:

3. Now, let's write a test to verify this behavior.

In this recipe, we will learn how to run the espresso test; in the next recipe, we will learn how to write an espresso test, so just bear with me until I explain how to write an espresso test in the later recipe because it is complicated and needs an entire recipe to do justice to it.

Here's an espresso test:

```
class MainActivityTest {
    @Rule
    @JvmField var activityRule: ActivityTestRule<MainActivity> =
ActivityTestRule(MainActivity::class.java)

    @Test
    fun testButtonBehaviour() {
        onView(withText("Hello World!"))
                .check(matches(isDisplayed()))
        onView(withId(button)).perform(click())
        onView(withText("Goodbye World!"))
                .check(matches(isDisplayed()))


    }

}
```

In the first line of the `testButtonBehaviour` method, we are checking whether **Hello World!** is appearing on screen. Then, we are performing a click operation on the button and finally checking whether **Goodbye World!** is appearing on the screen.

4. To run the preceding test, just right-click on the test class and select **Run MainActivityTest**:

5. Once you select that option, you'll be shown a dialog box that will ask which device you want to run the test on. You can either choose a real device or an emulator.

6. After that, you can see your tests running on your device (you will see the steps written in the code being performed on the device).

There's more...

If you run any instrumentation test, you'll note that it takes a lot of time to pass, even if it's a small test. In the current scenario, **test-driven development** (**TDD**) is becoming more and more popular, but its tests take a lot of time to execute; it's not a good way of using TDD. So the number of instrumentation tests should be kept to a minimum, and it's better to use a mocking framework, such as Mockito or Robolectric.

Wikipedia says that TDD is a software development process that relies on the repetition of a very short development cycle: Requirements are turned into very specific test cases, and then the software is improved to pass the new tests only.

Writing JUnit rules in Kotlin (@Rule)

Rules are a way to add functionalities that apply to all tests of the class. For example, `ExternalResource` executes the code before and after a test method. This can be used to set up a database, network, and filesystem connection before the test method, and can tear them down when the tests are complete. Of course, you can also do it using the `@Before` and `@After` annotations, but doing it with `ExternalResource` (as a JUnit rule) helps with code reuse.

Getting ready

I'll be using Android Studio 3.0 for coding.

How to do it...

In this recipe, we will be using `ExpectedException` as the JUnit rule because it helps the test declare that an exception is expected and also provides a way to clearly express the expected behavior. It is much more flexible than using the `@Test(expected= ...)` annotation because we can test specific error messages and custom fields.

In the following steps, we will learn how to write JUnit tests:

1. Let's first create a simple method that throws an exception. We will then write a test to test this method:

```
fun methodThrowsException() {
    throw IllegalArgumentException("Age must be integer")
}
```

2. Now, let's create a new rule of the `ExpectedException` class and write a test:

```
@Rule
var thrown = ExpectedException.none()

@Test
fun testExceptionFlow() {
    thrown.expect(IllegalArgumentException::class.java)
    thrown.expectMessage("Age must be integer")
    Utility.methodThrowsException()
}
```

3. If you run the preceding code, you will get an error:

```
org.junit.internal.runners.rules.ValidationError: The @Rule
'thrown' must be public.
```

4. The error is because JUnit allows the provision of rules through a test class field or a getter method. However, we don't have fields in Kotlin—we have properties; so what you've really annotated here is a property, not a field.

5. The easiest way around this problem is by adding the `@JvmField` annotation with `@Rule`:

```
@Rule @JvmField
var thrown = ExpectedException.none()
```

6. If you run the test now, it will pass.

How it works...

We are aware of the fact that Kotlin plays with properties rather than fields in Java. However, to provide compatibility with the Java language, `@JvmField` can be used to instruct the Kotlin compiler not to generate getters–setters for this property and expose it as a field.

However, there are a few restrictions when using annotations. We can't use them with the following:

- Private properties
- Properties with `open`, `override`, and `const` modifiers
- Delegated properties

Acceptance tests using Espresso Kotlin

Espresso is the most popular UI testing framework for Android. It was released by Google in 2013 and is the easiest to use of its kind. It provides support for complex things, such as ensuring that an activity is run before the tests are run, or waiting till the observed background tasks are completed. These things were hard to synchronize prior to Espresso, and UI testing was considered a difficult thing to do.

In this recipe, we will learn how to use Espresso to perform acceptance testing.

 Acceptance testing is a level of software testing where a system is tested for acceptability. The purpose of this test is to evaluate the system's compliance with the business requirements and assess whether it is acceptable for delivery.
Source: `http://softwaretestingfundamentals.com/`

Getting ready

We'll be using Android Studio 3.0 for our coding purposes. You can download the source code from `https://gitlab.com/aanandshekharroy/Anko-examples` and switch to the **5-instrumentation-tests** branch.

How to do it...

In Espresso, we mainly have three components:

- `ViewMatchers`: Allows you to find a view in the current view hierarchy. This can be done in various ways, such as searching by `id`, `name`, `child`, and so on. You can also use Hamcrest matchers, such as `containsString`.

- `ViewActions`: Allows you to perform actions on the views, such as clicking, typing, clearing text, and so on.

- `ViewAssertions`: Allows you to assert the state of a view that checks whether the condition under the view assertion passes.

Let's take a look at the following steps to understand acceptance testing using Espresso:

1. Here's an example of a text matcher (a text matcher matches the text; it's a part of `ViewMatchers`):

   ```
   onView(withId(R.id.textView)).check(matches(withText(not(containsString("Hello"))))));
   ```

2. Now we will create a simple test that will test whether clicking on the button changes the text from **Hello World!** to **Goodbye World!**:

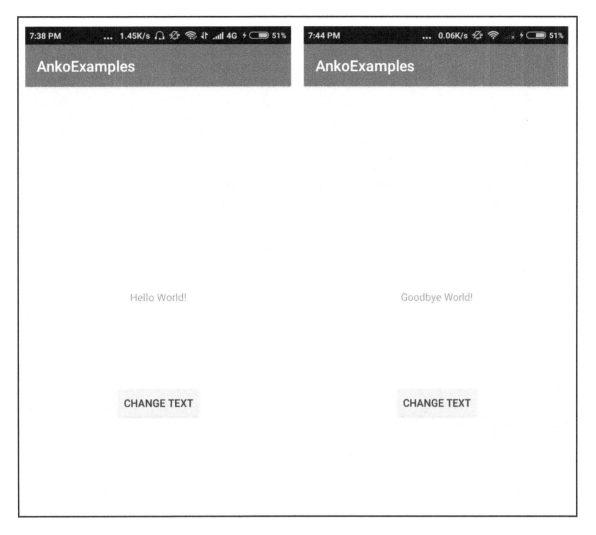

3. The following is an espresso test written to test the preceding functionality:

```
class MainActivityTest {
    @Rule
    @JvmField var activityRule: ActivityTestRule<MainActivity> =
ActivityTestRule(MainActivity::class.java)

    @Test
    fun testButtonBehaviour() {
        // Testing if the text is initially Hello World!
        onView(withText("Hello World!"))
```

```
                    .check(matches(isDisplayed()))
        onView(withId(button)).perform(click())
        // Testing if the text is initially Goodbye World!
        onView(withText("Goodbye World!"))
                    .check(matches(isDisplayed()))


    }


}
```

4. In the first line, we have created a rule that provides functional testing of a single activity. This will open the activity before the tests are run and close it once the tests are completed. The statements in the `testButtonBehaviour` method check the UI conditions, such as whether the text is initially **Hello World!** (first condition), then performs a click action on the button, and then finally checks whether the text is now **Goodbye World!**.

5. You can also get hold of the "God" object—that is, `Context`—using the instrumentation API:

```
var targetContext:Context =
InstrumentationRegistry.getTargetContext()
```

6. If you want to start the activity from the intent, you just need to provide `false` as the third argument in the constructor of `ActivityTestRule`; it can be used as shown:

```
@Rule
@JvmField var intentActivityRule: ActivityTestRule<MainActivity> =
ActivityTestRule(MainActivity::class.java,true,false)

@Test
fun testIntentLaunch(){
    val intent = Intent()
    intentActivityRule.launchActivity(intent)
    onView(withText("Hello World!"))
            .check(matches(isDisplayed()))
}
```

Another cool feature of Espresso is to record the interactions using **Record Espresso Test**. This records all the interactions that you make with the app, and can generate tests from it. To use that feature, follow these steps:

1. Go to **Run** on the toolbar and select **Record Espresso Test**:

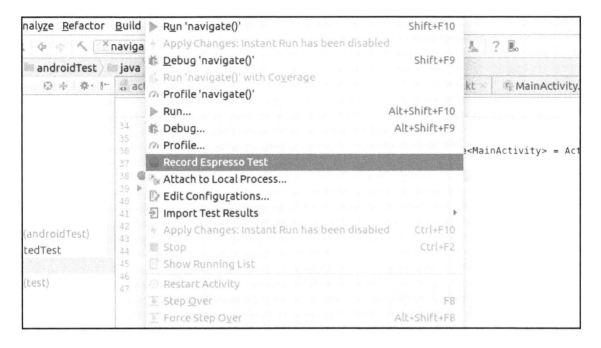

2. It will then open a dialog, on which you can see the recorded steps:

3. Clicking on **OK** will automatically generate the tests based on your interaction.

There's more...

Note that we have used the `@JvmField` annotation with the rule. The reason for this is extensively discussed in the *How to write JUnit rules in Kotlin (@Rule)* recipe.

Writing assertEquals in Kotlin

The `assertEquals` statement is widely used for testing code. It basically takes in two arguments—an expected value and an actual value—with an optional third argument message. If the expected value matches the actual value, the `assertEquals` passes—otherwise, it fails.

Using `assertEquals` with primitive types is straightforward, but if you want to use it with a custom object, you'll have to do a little more work. For example, the following `assertEquals` will not pass:

```
assertEquals(MyObj("abc"),MyObj("abc"))
```

In this recipe, we will learn how to write `assertEquals` statements.

Getting ready

We'll be using Android Studio 3.0 for our coding purposes. You can download the source code from `https://gitlab.com/aanandshekharroy/Anko-examples` and switch to the `5-instrumentation-tests` branch.

How to do it...

Let's go through the following steps to understand `assertEquals`:

1. In the following code, if you run the given `assertEquals`, it will not pass:

   ```
   assertEquals(MyObj("abc"),MyObj("abc"))
   ```

2. If you check the difference, it will tell you that they aren't equal because they are two different objects:

3. So we will need to override the `equals` method of the `MyObj` class, and we will check the following things:

   - Whether the other object references the same object—this can be done using the `===` operator, which checks the referential equality
   - Whether the object in question equals the other object's Java class
   - Whether the content of both objects is the same:

```
override fun equals(other: Any?): Boolean {
    if (this === other) return true
    if (other?.javaClass != javaClass) return false
    other as MyObj

    if (name != other.name) return false

    return true
}
```

Now, when you run `assertEquals` with two objects of the same content, it will pass smoothly.

14

Web Services with Kotlin

The following recipes will be covered in this chapter:

- How to run the application on Tomcat
- Setting up dependencies for building RESTful services
- How to create a REST controller
- Creating the Application class for Spring boot

Introduction

Kotlin has been eating up the Java world. It has already become a hit in Android Ecosystem which was dominated by Java and is welcomed with open arms everywhere. Kotlin is not limited to Android development and can be used to develop server-side, client-side web applications. One of the `use` cases that we will address in this chapter is creating web-services using Kotlin. Kotlin is 100% compatible with the JVM and so you can use any existing frameworks such as Spring Boot, Vert.x, or JSF for writing Java applications.

How to run the application on Tomcat

In this recipe, we will learn how to install, configure, and run the application on Tomcat in IntelliJ IDEA.

 Apache Tomcat, often referred to as Tomcat Server, is an open source Java Servlet Container developed by the Apache Software Foundation (ASF). Tomcat implements several Java EE specifications, including Java Servlet, JavaServer Pages (JSP), Java EL, and WebSocket, and provides a "pure Java" HTTP web server environment in which Java code can run. Source: Wikipedia

How to do it...

Now, let's follow the given steps to run the application on Tomcat:

1. First, you need to download the Tomcat from `http://tomcat.apache.org/download-80.cgi`.

2. The downloaded file will be a compressed file, and you can extract it with:

   ```
   tar xvzf apache-tomcat-8.0.9.tar.gz
   ```

3. Next, you need to move it from the downloaded folder to the proper location, at:

   ```
   mv apache-tomcat-8.0.9 /opt/tomcat
   ```

4. You also need to check whether you have JDK set up on your system. You can do that by typing in the following command:

   ```
   java -version
   ```

5. If you see **The program 'java' can be found in the following packages:**, it means you need to install JDK. You can do it with the following:

   ```
   sudo apt-get install openjdk-7-jdk
   ```

6. After that, add the following lines to the end of the `.bashrc` file:

```
export JAVA_HOME=/usr/lib/jvm/java-7-openjdk-amd64
export CATALINA_HOME=/opt/tomcat
```

7. Simply save and exit `.bashrc`, and then make the changes effective by running the following command:

```
. ~/.bashrc
```

8. Tomcat and Java should now be installed and configured on your server. To activate Tomcat, run the following script:

```
$CATALINA_HOME/bin/startup.sh
```

You should get a result similar to the following:

```
Using CATALINA_BASE: /opt/tomcat
Using CATALINA_HOME: /opt/tomcat
Using CATALINA_TMPDIR: /opt/tomcat/temp
Using JRE_HOME: /usr/lib/jvm/java-7-openjdk-amd64/
Using CLASSPATH:
/opt/tomcat/bin/bootstrap.jar:/opt/tomcat/bin/tomcat-juli.jar
Tomcat started.
```

9. Open `http://127.0.0.1:8080` to check if it's working.

10. Now, you'll need IntelliJ IDEA's ultimate edition to be able to use Tomcat in IntelliJ; community edition doesn't provide support for Java EE application.

11. In order to run the application, we need the corresponding WAR(s) for deploying, which you can do just by adding the following lines in the terminal:

```
gradle war
```

12. You need to go to **Run** | **Edit configuration** and add **Tomcat**:

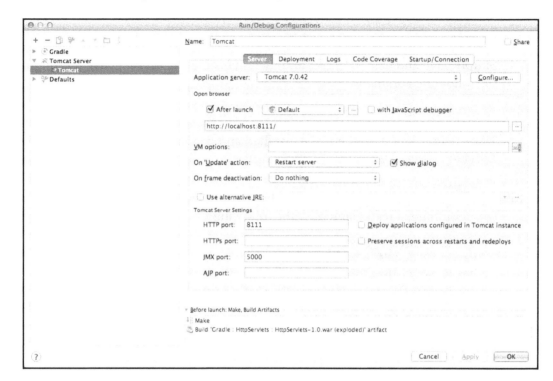

13. Now if you move to your localhost server, you can see the application hosted there.

For instructions on Windows installation of Tomcat, refer to `https://www.ntu.edu.sg/home/ehchua/programming/howto/Tomcat_HowTo.html`.

Setting up dependencies for building RESTful services

In this recipe, we will lay the foundation for developing the RESTful service. We will see how to set up dependencies and run our first SpringBoot web application. SpringBoot provides great support for Kotlin, which makes it easy to work with Kotlin. So let's get started.

Getting ready

We will be using IntelliJ IDEA and Gradle build system. If you don't have that, you can get it from `https://www.jetbrains.com/idea/`.

How to do it...

Let's follow the given steps to set up the dependencies for building RESTful services:

1. First, we will create a new project in IntelliJ IDE. We will be using the Gradle build system for maintaining dependency, so create a `Gradle` project:

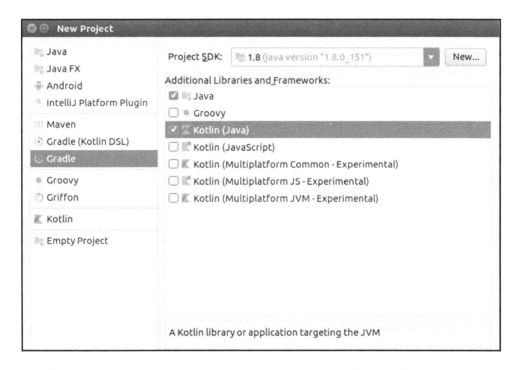

2. When you have created the project, just add the following lines to your `build.gradle` file. These lines of code contain spring-boot dependencies that we will need to develop the web app:

```
buildscript {
    ext.kotlin_version = '1.1.60' // Required for Kotlin
integration
    ext.spring_boot_version = '1.5.4.RELEASE'
```

```
        repositories {
            jcenter()
        }
        dependencies {
            classpath "org.jetbrains.kotlin:kotlin-gradle-
plugin:$kotlin_version" // Required for Kotlin integration
            classpath "org.jetbrains.kotlin:kotlin-
allopen:$kotlin_version" // See
https://kotlinlang.org/docs/reference/compiler-plugins.html#kotlin-
spring-compiler-plugin
            classpath "org.springframework.boot:spring-boot-gradle-
plugin:$spring_boot_version"
        }
}

apply plugin: 'kotlin' // Required for Kotlin integration
apply plugin: "kotlin-spring" // See
https://kotlinlang.org/docs/reference/compiler-plugins.html#kotlin-
spring-compiler-plugin
apply plugin: 'org.springframework.boot'

jar {
    baseName = 'gs-rest-service'
    version = '0.1.0'
}
sourceSets {
    main.java.srcDirs += 'src/main/kotlin'
}

repositories {
    jcenter()
}

dependencies {
    compile "org.jetbrains.kotlin:kotlin-stdlib:$kotlin_version" //
Required for Kotlin integration
    compile 'org.springframework.boot:spring-boot-starter-web'
    testCompile('org.springframework.boot:spring-boot-starter-
test')
}
```

3. Let's now create an `App.kt` file in the following directory hierarchy:

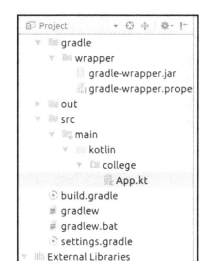

It is important to keep the `App.kt` file in a package (we've used the `college` package). Otherwise, you will get an error that says the following:

```
** WARNING ** : Your ApplicationContext is unlikely to start due to
a `@ComponentScan` of the default package.
```

The reason for this error is that if you don't include a package declaration, it considers it a "default package," which is discouraged and avoided.

4. Now, let's try to run the `App.kt` class. We will put the following code to test if it's running:

```
@SpringBootApplication
open class App {
}

fun main(args: Array<String>) {
    SpringApplication.run(App::class.java, *args)
}
```

5. Now run the project; if everything goes well, you will see output with the following line at the end:

```
Started AppKt in 5.875 seconds (JVM running for 6.445)
```

6. We now have our application running on our embedded Tomcat server. If you go to `http://localhost:8080`, you will see an error as follows:

Whitelabel Error Page

This application has no explicit mapping for /error, so you are seeing this as a fallback.

Thu Nov 16 09:56:54 IST 2017
There was an unexpected error (type=Not Found, status=404).
No message available

7. The preceding error is `404 error` and the reason for that is we haven't told our application to do anything when a user is on the `/` path.

How to create a REST controller

In the previous recipe, we learned how to set up dependencies for creating RESTful services. Finally, we launched our backend on the `http://localhost:8080` endpoint but got `404 error` as our application wasn't configured to handle requests at that path (`/`). We will start from that point and learn how to create a REST controller. Let's get started!

Getting ready

We will be using IntelliJ IDE for coding purposes. For setting up of the environment, refer to the previous recipe. You can also find the source in the repository at `https://gitlab.com/aanandshekharroy/kotlin-webservices`.

How to do it...

In this recipe, we will create a REST controller that will fetch us information about students in a college. We will be using an in-memory database using a list to keep things simple:

1. Let's first create a `Student` class having a name and roll number properties:

```
package college

class Student() {
    lateinit var roll_number: String
    lateinit var name: String
    constructor(
            roll_number: String,
            name: String): this() {
        this.roll_number = roll_number
        this.name = name
    }
}
```

2. Next, we will create the `StudentDatabase` endpoint, which will act as a database for the application:

```
@Component
class StudentDatabase {
    private val students = mutableListOf<Student>()
}
```

Note that we have annotated the `StudentDatabase` class with `@Component`, which means its lifecycle will be controlled by Spring (because we want it to act as a database for our application).

3. We also need a `@PostConstruct` annotation, because it's an in-memory database that is destroyed when the application closes. So we would like to have a filled database whenever the application launches. So we will create an `init` method, which will add a few items into the "database" at startup time:

```
@PostConstruct
private fun init() {
    students.add(Student("2013001","Aanand Shekhar Roy"))
    students.add(Student("2013165","Rashi Karanpuria"))
}
```

4. Now, we will create a few other methods that will help us deal with our database:

   - getStudent: Gets the list of students present in our database:

     ```
     fun getStudents()=students
     ```

   - addStudent: This method will add a student to our database:

     ```
     fun addStudent(student: Student): Boolean {
         students.add(student)
         return true
     }
     ```

5. Now let's put this database to use. We will be creating a REST controller that will handle the request. We will create a StudentController and annotate it with @RestController. Using @RestController is simple, and it's the preferred method for creating MVC RESTful web services.

6. Once created, we need to provide our database using Spring dependency injection, for which we will need the @Autowired annotation. Here's how our StudentController looks:

   ```
   @RestController
   class StudentController {
       @Autowired
       private lateinit var database: StudentDatabase
   }
   ```

7. Now we will set our response to the / path. We will show the list of students in our database. For that, we will simply create a method that lists out students. We will need to annotate it with @RequestMapping and provide parameters such as path and request method (GET, POST, and such):

   ```
   @RequestMapping("", method = arrayOf(RequestMethod.GET))
   fun students() = database.getStudents()
   ```

8. This is what our controller looks like now. It is a simple REST controller:

   ```
   package college

   import org.springframework.beans.factory.annotation.Autowired
   import org.springframework.web.bind.annotation.RequestMapping
   import org.springframework.web.bind.annotation.RequestMethod
   import org.springframework.web.bind.annotation.RestController
   ```

```
@RestController
class StudentController {
    @Autowired
    private lateinit var database: StudentDatabase
    @RequestMapping("", method = arrayOf(RequestMethod.GET))
    fun students() = database.getStudents()
}
```

9. Now when you restart the server and go to `http://localhost:8080`, we will see the response as follows:

As you can see, Spring is intelligent enough to provide the response in the JSON format, which makes it easy to design APIs.

10. Now let's try to create another endpoint that will fetch a student's details from a roll number:

```
@GetMapping("/student/{roll_number}")
fun studentWithRollNumber( @PathVariable("roll_number")
roll_number:String) =
    database.getStudentWithRollNumber(roll_number)
```

11. Now, if you try the `http://localhost:8080/student/2013001` endpoint, you will see the given output:

```
{"roll_number":"2013001","name":"Aanand Shekhar Roy"}
```

12. Next, we will try to add a student to the database. We will be doing it via the `POST` method:

```
@RequestMapping("/add", method = arrayOf(RequestMethod.POST))
fun addStudent(@RequestBody student: Student) =
        if (database.addStudent(student)) student
        else throw Exception("Something went wrong")
```

There's more...

So far, our server has been dependent on IDE. We would definitely want to make it independent of IDE. Thanks to Gradle, it is very easy to create a runnable JAR just with the following:

```
./gradlew clean bootRepackage
```

The preceding command is platform independent and uses the Gradle build system to build the application. Now, you just need to type the mentioned command to run it:

```
java -jar build/libs/gs-rest-service-0.1.0.jar
```

You can then see the following output as before:

```
Started AppKt in 4.858 seconds (JVM running for 5.548)
```

This means your server is running successfully.

Creating the Application class for Spring Boot

The `SpringApplication` class is used to bootstrap our application. We've used it in the previous recipes; we will see how to create the `Application` class for Spring Boot in this recipe.

Getting ready

We will be using IntelliJ IDE for coding purposes. To set up the environment, read previous recipes, especially the *Setting up dependencies for building RESTful services* recipe.

How to do it...

If you've used Spring Boot before, you must be familiar with using `@Configuration`, `@EnableAutoConfiguration`, and `@ComponentScan` in your main class. These were used so frequently that Spring Boot provides a convenient `@SpringBootApplication` alternative. The Spring Boot looks for the `public static main` method, and we will use a top-level function outside the `Application` class.

If you noted, while setting up the dependencies, we used the `kotlin-spring` plugin, hence we don't need to make the `Application` class open.

Here's an example of the Spring Boot application:

```
package college

import org.springframework.boot.SpringApplication
import org.springframework.boot.autoconfigure.SpringBootApplication

@SpringBootApplication
class Application
fun main(args: Array<String>) {
    SpringApplication.run(Application::class.java, *args)
}
```

The Spring Boot application executes the static `run()` method, which takes two parameters and starts a autoconfigured Tomcat web server when Spring application is started.

When everything is set, you can start the application by executing the following command:

```
./gradlew bootRun
```

If everything goes well, you will see the following output in the console:

This is along with the last message—**Started AppKt in xxx seconds**. This means that your application is up and running.

In order to run it as an independent server, you need to create a JAR and then you can execute as follows:

```
./gradlew clean bootRepackage
```

Now, to run it, you just need to type the following command:

```
java -jar build/libs/gs-rest-service-0.1.0.jar
```

Other Books You May Enjoy

If you enjoyed this book, you may be interested in these other books by Packt:

Reactive Programming in Kotlin
Rivu Chakraborty

ISBN: 978-1-78847-302-6

- Learn about reactive programming paradigms and how reactive programming can improve your existing projects
- Gain in-depth knowledge in RxKotlin 2.0 and the ReactiveX Framework
- Use RxKotlin with Android
- Create your own custom operators in RxKotlin
- Use Spring Framework 5.0 with Kotlin
- Use the reactor-Kotlin extension
- Build Rest APIs with Spring,Hibernate, and RxKotlin
- Use testSubscriber to test RxKotlin applications
- Use backpressure management and Flowables

Programming Kotlin
Stephen Samuel, Stefan Bocutiu

ISBN: 978-1-78712-636-7

- Use new features to write structured and readable object-oriented code
- Find out how to use lambdas and higher-order functions to write clean, reusable, and simple code
- Write unit tests and integrate Kotlin tests with Java code in a transitioning code base
- Write real-world production code in Kotlin in the style of microservices
- Leverage Kotlin's extensions to the Java collections library
- Use destructuring expressions and find out how to write your own
- Write code that avoids null pointer errors and see how Java-nullable code can integrate with features in a Kotlin code base
- Discover how to write functions in Kotlin, see the new features available, and extend existing libraries
- Learn to write an algebraic data types and figure out when they should be used

Leave a review - let other readers know what you think

Please share your thoughts on this book with others by leaving a review on the site that you bought it from. If you purchased the book from Amazon, please leave us an honest review on this book's Amazon page. This is vital so that other potential readers can see and use your unbiased opinion to make purchasing decisions, we can understand what our customers think about our products, and our authors can see your feedback on the title that they have worked with Packt to create. It will only take a few minutes of your time, but is valuable to other potential customers, our authors, and Packt. Thank you!

Index